PHYSICS
CURIOSITIES, ODDITIES, AND NOVELTIES

PHYSICS
CURIOSITIES, ODDITIES, AND NOVELTIES

JOHN KIMBALL

CRC Press
Taylor & Francis Group
Boca Raton London New York

CRC Press is an imprint of the
Taylor & Francis Group, an **informa** business

CRC Press
Taylor & Francis Group
6000 Broken Sound Parkway NW, Suite 300
Boca Raton, FL 33487-2742

International Standard Book Number-13: 978-1-4665-7635-3 (Paperback)

Library of Congress Cataloging-in-Publication Data

Kimball, John, 1942- author.
  Physics curiosities, oddities, and novelties / John Kimball.
    pages cm
  Includes bibliographical references and index.
  ISBN 978-1-4665-7635-3
  1. Physics--Popular works. I. Title.

QC24.5.K56 2015
530--dc23   2014046008

Visit the Taylor & Francis Web site at
http://www.taylorandfrancis.com

and the CRC Press Web site at
http://www.crcpress.com

Friends and collaborators I have known over many years have contributed to the ideas contained in this book. Special thanks for extra help are due to Oleg Lunin, T. S. Kuan, and David Liguori at the University at Albany.

Contents

Preface

Physics is permeated with mind-boggling concepts. Indeed, why does so much of physics seem tired or incomprehensible? It does not help that the "old testament" of physics, Newton's *Principia*, really is old. Even "modern physics" was established a century ago. The tedium frequently associated with physics often takes the form of long mathematical calculations—it is the price one pays to obtain precise results. The idea that an object can be both a wave and a particle is pretty strange. One is bound to be confused by the relativistic idea that clocks cannot be trusted.

This book highlights unusual aspects of physics while giving a new twist to some fundamental concepts. The goal has been to add a little zip to classical physics and make modern physics at least plausible. This requires bending the rules. Not all the standard textbook topics are covered. I have chosen examples that reveal interesting and unusual ideas. Mathematics is minimized. A bit of gossip about eccentric scientists has been included. Perplexing questions are raised that lack satisfying answers. The following are some example topics

- The ancient Greeks determined the size of the moon, but even Newton had trouble determining its mass.
- An electric motor is an electric generator run in reverse.

- Chaos and quantum mechanics challenge the determinism of classical physics, but they do not destroy it.
- Nuclear physics tells us that the Shroud of Turin is not as old as it should be.
- Special relativity means rapidly moving objects appear to shrink. But even at top speed, a camel cannot pass through the eye of a needle.
- Time travel that violates causality is not possible.
- Schrodinger's cat may be both dead and alive, and there may be two of each one of us to observe the two possibilities.
- Feynman diagrams can deceive one into thinking particle physics is simple. (It is really complicated.)
- The basic laws of thermodynamics can also appear simple, even though the applications are often surprising.
- Many physics Nobel Prize winners are agnostics or atheists, but the religious Nobel Laureates are more interesting.

I hope that curious people of many backgrounds find things of interest in this book. Many people fully capable of understanding abstract physical ideas do not like equations. For example, those immersed in physics never forget that $\vec{p}$ means momentum, and they know exactly what momentum denotes. But, this knowledge comes from years of conditioning. Typical readers lack this training, so care has been taken to explain the meanings carefully. However, a book about physics that completely abandons math is going a step too far. It would be a travesty to pretend that $\vec{F} = m\vec{a}$ or $E = mc^2$ are not worth knowing. My goal has been to summarize many interesting aspects of physics using only essential formulas. Some of the most familiar ones have been written in standard form, but if the shorthand notation is not necessary, equations have been written out in words for greater clarity.

I believe it is possible to provide insight into physics while minimizing jargon. Nonspecialists and beginning students often ask the most difficult questions. They deserve answers. I think this can be accomplished while avoiding specialized lingo and without talking down to the reader.

About the Author

John Kimball taught physics at Louisiana State University from 1971 through 1978. After brief periods of teaching at the University of California, he taught physics from 1981 at the University at Albany, New York, slipping gradually into retirement by 2012. His research has largely focused on condensed matter physics. Interesting problems addressed by Dr. Kimball and colleagues include high-energy electron-positron pair production in crystals, exact solutions of a kinetic model of magnetism, relations between chaos and symmetry, and studies of magnetic impurities.

Dr. Kimball's undergraduate education was at the University of Minnesota and his graduate work was at the University of Chicago, working with Professor Leo Falicov. Postdoctoral work was done at the University of Pennsylvania under the direction of Dr. Robert Schrieffer. Many thanks are due Professors Falicov and Schrieffer and to the universities in Minnesota, Chicago, Pennsylvania, Louisiana, and Albany for making life as a physicist so rewarding.

John Kimball is an enthusiastic sailor with limited sailing skills and author of the book *The Physics of Sailing*, also published by Taylor & Francis.

1

Newton and Mechanics

1.1 Introduction

Isaac Newton (Figure 1.1) revolutionized science when he introduced the world to his equation $\vec{F} = m\vec{a}$ and the "universal law of gravity." These two concepts explain why all objects fall at the same rate. They describe planetary orbits, the motion of our moon, and much more. Newton gave physics a revolutionary theory that was capable of both explaining and predicting phenomena on Earth and in the sky.

1.2 Newton's Equation

Newton combined observations of motion with mathematical skill to develop his basic equation of mechanics: $\vec{F} = m\vec{a}$.

Isaak Newton

Figure 1.1 Isaak Newton: a complimentary image of the great man.

1.2.1 Data for Newton: Falling

Objects dropped from the same height take the same time to hit the ground. This remarkably simple observation is the most important and easiest experiment in physics. Drop a big rock and a little rock and see what happens. One should really be a little more careful. All objects fall identically only in a vacuum, where the annoying drag and buoyancy of air can be ignored. For a typical rock-dropping experiment, Earth's atmosphere has a negligible effect. Comparing a falling feather with a falling rock only obscures the physics.

Newton's ideas were not germinated by an apple falling on his head, and he had no need to perform rock-dropping experiments. The clear-thinking Galileo and others had already determined the experimental facts about falling that are so brilliantly explained by Newton's theories. Newton was neither the first nor the last person to have rightly said: "If I have seen further, it is by standing on the shoulders of giants." (Actually, Newton's use of this often-repeated idea may not have been an expression of modesty. He may have been poking fun at the unusually short Robert Hook, who had "issues" with Newton.)

Surely, the simple observation that all objects take the same time to fall must have been well known since ancient times. But wait—Aristotle said heavy things fall faster than light ones. He postulated that the motion of a substance depended on its nature. An extreme example was "fire," whose nature caused it to rise rather than fall. It is remarkable how confusing something as simple as falling can appear when one is initially misguided. Aristotle should have performed the big-rock and small-rock experiment, but he (and some other Greeks of his time) trusted their cleverness more than mundane trial and error. The most famous criticism of Aristotle's reason-based rather than experiment-based science was Bertrand Russell's comment:

> Aristotle maintained that women have fewer teeth than men; although he was twice married, it never occurred to him to verify this statement by examining his wives' mouths.

Of course it is unreasonable to expect even a brilliant observer to count the teeth of his spouse. The fault lies in formulating a theory without the benefit of an experiment. Some of Aristotle's wrong ideas had amazing longevities. It may seem strange that Galileo needed to argue so strenuously and carefully that all objects fall at the same rate, but he was fighting against nearly two millennia of "wisdom of the ancients." Aristotle's view of nature was an important source of this ancient wisdom. There is a lesson here, well-stated by Galileo:

> In questions of science, the authority of a thousand is not worth the humble reasoning of a single individual.

The lasting effects of Aristotle's ideas are all the more surprising in view of the many brilliant contributions to science and engineering of other ancient Greeks. Pythagoras, who lived about 200 years earlier than Aristotle, is the best-known example. His ideas on acoustics were based on experiments rather than speculation.

1.2.2 Newton's Tools: Mathematics

Physics uses math. There is no point in pretending otherwise. It is unfortunate that the world divides into two camps, famously described by Sir C. P. Snow in a 1959 lecture "The Two Cultures." The first culture, which cannot abide mathematics, has trouble seeing merit in the second culture. People in the second culture know there is no escape from mathematics if one wishes to know even a little about physics.

Many famous physicists (as well as those not so famous) have marveled at the inseparable connection of science and math. For example, Richard Feynman, one of the creators of quantum electrodynamics, said:

> To those who do not know mathematics it is difficult to get across a real feeling as to the beauty, the deepest beauty of nature.

Eugene Wigner made important contributions to many areas of modern physics, especially the consequences of symmetries. Today, he is probably best known for his 1960 article, "The Unreasonable Effectiveness of Mathematics in the Natural Sciences." He marveled at the magic that makes mathematics so successful. However, Wigner's optimism was limited. He frequently commented that humans may not have the intellectual capability to deal with the most fundamental questions. He wondered if our inability to understand the universe was comparable to a dog's inability to master calculus. Others, notably the famous particle theorist Steven Weinberg, feel that humans will eventually achieve an ultimate theory that explains everything.

Dr. Weinberg's optimism is emotionally appealing, but I agree with Wigner's dog-learning-calculus point of view.

1.2.3 Newton's Equation

This is the most important equation in physics:

$$\vec{F} = m\vec{a}$$

What does this mean? The equation is more than a bit of algebra. The letters are codes for precisely defined quantities. Without the shorthand codes, the wordy version of Newton's equation is

$$Force = (Mass) \times (Acceleration)$$

To make sense of this equation, begin with an oversimplification: *Mass* is the amount of material. For motion along a straight line, *acceleration* is the rate at which speed changes. *Force* is the "push" on the object. This suggests that Newton is only telling us something trivial—push on something to make it move faster.

But, any suggestion of triviality insults Newton. Newton's equation is more than conceptual. It gives exact results for real-world situations. The arrows over the force $\vec{F}$ and the acceleration $\vec{a}$ mean the force and acceleration are in the same direction. In situations where direction is understood, the superscript arrows are often suppressed.

1.2.4 Newton's Laws

Newton also put his equation into words, as expressed in his first two laws, published in *Philosophiæ Naturalis Principia Mathematica* (1687). In modern language (and not in Latin), they are as follows:

First law: When there is no force, an object at rest stays at rest. When there is no force, a moving object continues to move at the same speed with no change in direction.

Second law: When there is a force, the acceleration $\vec{a}$ of a body is parallel and directly proportional to the net force $\vec{F}$ acting on the body and is inversely proportional to the mass m of the body.

Pragmatically, everything stated in Newton's first and second laws is a consequence of $\vec{F} = m\vec{a}$. It saves time and brain capacity to forget the first two laws and just remember this single compact equation. In this sense, Newton's first law is just a special case of the second law. However, in Newton's time, all three of his laws represented revolutionary ideas.

From a philosophical rather than pragmatic view, Newton's first law is an assertion that properly chosen coordinates of space and time are the foundations of physics. Not all space-time coordinates are consistent with Newton's equation. Someone riding in an accelerating car will see objects moving past at increasing speed even though the objects are not subject to forces. Similarly, a rotating observer would think objects subjected to no force travel along curved paths. Distinguishing accelerating coordinates from forces is difficult, as noted in Section 1.2.9. However, it is easier to identify rotating coordinates. "Newton's bucket" is the key. If a rotating bucket is partially filled with water, the water will push toward the sides of the bucket and the surface will not be flat. Thus, absolute rotation can be observed without reference to any outside objects.

To appreciate the universal applicability and utility of Newton's equation $F = ma$, one must carefully define *mass, acceleration,* and *force.* Mass and acceleration are described first. Gravity is the most important force described by Newton. It is described in Section 1.3.

1.2.5 Mass m

Mass and weight are not the same. In 1971, astronaut Alan Shepard managed to drive a golf ball a remarkably long distance. Alan performed this trick by golfing on the moon, where objects *weigh* less. But, the mass m of a golf ball (or any other object) is the same no matter where it is located. Mass is a measure of the amount of material, and that is not the same as weight. It is convenient to measure mass in kilograms. On Earth, 1 kilogram weighs about 2.2 pounds. On the moon, the same kilogram weighs less than 6 ounces. An astronaut floating through space may be "weightless," but the astronaut is not "massless."

1.2.6 Acceleration ā and Acceleration of Gravity g

Acceleration is not speed. It is the rate that speed changes. For a falling rock starting at rest, the longer the acceleration takes place, the faster the rock moves because speed is acceleration multiplied by time. This can be expressed by another equation. The speed v after a time t is

$$v = at$$

When direction is important, the equation is $\vec{v} = \vec{a}t$, and a speed with its direction indicated is usually called *velocity*.

In the spirit of science, it is reasonable to ask why all heavy objects dropped from the same height take the same time to hit the ground. After all, it was recklessly claimed that this result was the simplest and most important experiment in physics. Surely, the most important experiment deserves an explanation. On a superficial level, Newton's equation provides the answer: All dropped objects (on Earth) take the same time to fall because they have the same acceleration.

This is not really an explanation. One can ask why the acceleration should be the same for every object, whether made of gold or lead. Newton's universal law of gravity (Section 1.3) yields the constant acceleration. Of course, one could still question the reason for gravity. In physics, the answer to one question often leads to another, often more difficult, problem.

If Galileo had dropped a large rock and a small rock from the Leaning Tower of Pisa (he probably did not) and if he had sufficiently accurate timers (he did not), he could have measured the constant acceleration suggested by Figure 1.2. Lacking accurate timers, Galileo

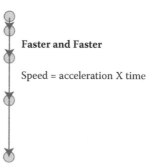

Figure 1.2 An accelerating (falling) object increases its speed with time. The falling ball positions are shown at 0, 1, 2, 3, and 4 seconds.

experimented with balls rolling down inclined planes. He showed that sphere rolling is also characterized by a constant, but smaller, acceleration. Galileo probably initially used the periodicity of music as his timer. Singing a sixteenth-century equivalent to "Tramp, tramp, tramp the boys are marching," he would mark the position of his rolling ball at each beat. Measuring the mark positions showed that the total distance traveled between beats was proportional to the time.

It was only a few years after Galileo published his ideas on constant acceleration that careful measurements of gravitational acceleration were done by dropping rocks and using a pendulum as a timer. (The idea that a pendulum makes a good timing device was another of Galileo's discoveries.) Measurement of the rock-dropping experiment with modern timers tells us that all objects dropped on Earth have an acceleration called g (presumably named for "gravity") and $g = 9.8$ meter / (second)^2. Using $v = at$ with $a = g$ means any object falling for 1 second (big rock or small) obtains a speed of 9.8 meters/second. At the end of 2 seconds, the speed will be doubled to 19.6 meters/second.

1.2.7 Ballistics

The constant acceleration of gravity determines the paths of objects flying through the air—provided air resistance can be ignored. When David sent his stone toward Goliath, how did he compensate for gravity? What would be the best direction to aim for maximum range? This is just one example of how the trajectories of flying objects have interested people for centuries. Scientists have long been attracted to this problem for intellectual, as well as military-financial, considerations. Archimedes designed weaponry, including a "steam cannon," used during the Roman siege of Syracuse in 214 BC. Sadly, Archimedes's inventions were not sufficient to prevent a Roman victory and his subsequent death at the hands of a Roman soldier.

The trajectories of flying objects are also of central importance in many sports. What direction should a batter hit a baseball when trying for a home run? Similar questions apply for golf balls and the big iron spheres of the shot put. Physics provides a single unifying answer to these questions. The results of Figure 1.3 apply to any flying object when the resistance of air can be ignored. The distance traveled depends only on the initial speed, the angle above horizontal, and the acceleration of gravity.

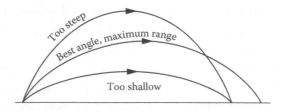

Figure 1.3 The path of three objects thrown at the same speed but in different directions. Starting at a 45-degree angle gives greater range than either a larger 60-degree angle or a smaller 30-degree angle. Air resistance is ignored.

Achieving maximum range boils down to a competition between two considerations. Throwing a rock straight up is clearly unwise because it will make no progress toward the target, and it will fall back on your head. Throwing horizontally is nearly as unproductive because the rock quickly hits the ground. Simplicity suggests a 45-degree angle above the horizontal for maximum range. In this case, simplicity is correct. At this compromise angle, the projectile stays in the air for a relatively long time because of its initial upward speed.

Shoot an arrow twice as fast and it will go four times as far (neglecting air resistance). The range increases with the squared speed because faster means more distance per unit time and more time in the air.

A golf ball on the moon goes six times farther because the lunar acceleration of gravity is six times smaller. In practice, Alan Shepard's best golf shot on the moon was not six times his Earth range. A cumbersome pressurized space suit cramped his style. However, the absence of a lunar atmosphere meant that no matter how carelessly Alan hit the ball, he had no need to worry about a slice or a hook. Without an atmosphere, the golf ball's path is guaranteed to follow a parabolic curve like one of those in Figure 1.3.

1.2.8 Lift and Drag

In the absence of air resistance (called *drag*), all dropped objects (even feathers) would accelerate identically. The speed of sound is 340 meters/second. Without drag, a falling human or marshmallow would acquire a supersonic speed after 35 seconds. Of course, this does not happen even if the object is dropped from 6 kilometers so it has time to fall for 35 seconds. Drag becomes increasingly important and acceleration stops as the falling object acquires its "terminal speed." This speed depends on size, shape, and mass. When your parachute collapses, your terminal speed increases dramatically. The terminal speed of a falling ping-pong ball is less than 10 meters/second or no faster than a sprinter can run. The terminal speed of a dropped iron shot used in shot put is about 15 times larger or about half the speed of sound.

Spinning balls and the wings of birds and airplanes are subject to an additional sideways force called *lift*. Table tennis experts use the drag and lift of air to their advantage. Topspin makes the ball drop. Drag and lift change ballistics. The maximum range is no longer achieved by a launch angle of 45 degrees. Golfers and baseball players know this. For the discus (essentially a heavy-duty Frisbee), experience has shown that a launch at 35 degrees above horizontal maximizes the range. Quantitative estimates of the drag force are given in Chapter 2.

1.2.9 Gravity and Acceleration

Is acceleration on Earth always $g = 9.8$ meters $/$ (second)^2? There are small variations (see Section 1.3.5), but the basic answer is yes, provided the experiment is done in an "inertial" coordinate system. Inertial does not mean fixed in space; it means moving at a constant velocity. For example, if the rock-dropping experiment is done on a moving train, the result is exactly the same. A test experiment done moving at any constant velocity in any direction, even up or down, yields the same result because these are all inertial coordinates. Experiments done in a noninertial accelerating coordinate system are different, as Fred notices.

Fred is placed in a box without windows. His location and motion are unknown. He releases a rock, and it does not move. What does this mean?

1. *Fred's first guess*: The box is far out in space where there is no gravity. Newton's equation $\vec{F} = m\vec{a}$ remains valid because no force means no acceleration.
2. *Fred's second guess*: Fred, his box, and his rock are accelerating toward Earth with the acceleration of Earth's gravity g. Because the observer is accelerating, the accelerating rock appears to be stationary. Force and acceleration are present, but Fred notices neither.

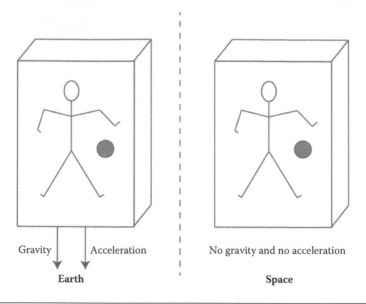

Figure 1.4 Fred notices that his ball is not moving. Is he falling, or is there no gravity?

There is no easy way to tell which of Fred's options shown in Figure 1.4 is correct. This can appear as circular reasoning and a flaw in Newtonian physics. Newton's equation requires inertial coordinates. But, inertial coordinates are defined as coordinates for which Newton's equation is valid. There is another way to state this ambiguity. It is nearly impossible to distinguish between acceleration and gravity. This is a key to general relativity.

1.2.10 Acceleration Examples

John Stapp was called "the fastest man alive." In 1962, he was strapped into a rocket sled that accelerated from 0 to 632 miles per hour (283 meters/second) in only 5 seconds. The formula $v = at$ means the average acceleration is the final speed divided by the time needed to achieve that speed. For John's sled ride, this was 57 meters/second^2. Often, accelerations are expressed as multiples of the acceleration of gravity. In gravity units John's acceleration was about $5.9g$.

Newton's equation says large accelerations are accompanied by large forces. For humans, the forces associated with really large accelerations are unpleasant and dangerous, in many cases because of diminished oxygen supplied to the brain. After undergoing many grueling experiments, some with accelerations greater than 5.9g, Stapp's vision was affected by broken blood vessels in his eyes, but he lived to age 89.

Research on human tolerance of acceleration has continued since the time of John Stapp. Jets and rockets can be constructed that produce more acceleration than jet pilots or astronauts can tolerate, so the human limits are important. We now know that people can lose their color vision (temporarily) at around 4g. Prolonged accelerations greater than about 5g can cause hallucinations and blackouts.

Manned rockets are designed to consider the comfort of their passengers as well as the structural integrity of the ship, so they typically restrict their accelerations to around 3g or less. Even though rockets limit their acceleration, they achieve high speeds by accelerating for a long time. To achieve Earth orbit, a rocket must reach a speed of 7890 meters/second. If the mean acceleration is 20 meters/second^2, acceleration should continue for almost 400 seconds to obtain orbital speed. This estimate of 400 seconds is approximate because the rocket's acceleration varies considerably along its path, the path is not a straight line, and Earth's rotation means the rocket is not really starting from rest.

1.2.11 Force $\vec{F}$

To honor Newton, forces are measured in "newtons." The most familiar force is gravity, pulling everything down. It is finally time to apply Newton's formula. A dropped 1-kilogram mass falls with an acceleration of 9.8 (meters)/(second)^2. For this example, $F = ma$ means the gravitational force on 1 kilogram must be 9.8 newtons. Because the acceleration of gravity is the same for all objects, the force is always proportional to the mass. Multiplying the gravitational acceleration and the mass of an elephant (5000 kilograms) gives an elephant force of 49,000 newtons.

If something is not falling, it is not accelerating. Newton tells us that zero acceleration means zero force. So, what happened to gravity for things just sitting around? The answer is that there is an equal and opposite force pushing the object up. Gravity pulling the elephant down is canceled by the ground's upward force of 49,000 newtons on the bottom of the elephant's feet. It can be painfully obvious to any weightlifter that the larger the mass, the more upward force is needed to keep it from falling. You could tell a weightlifter that $\vec{F} = m\vec{a}$ means lifting 100 kilograms requires 980 newtons to counter gravity. The weightlifter might not find this useful.

1.2.12 *Fast Cars and Animals*

Expensive racy cars, like the Chevrolet Corvette (perhaps not the car in Figure 1.5), can achieve a speed of 60 miles per hour in less than 3.5 seconds. Changing miles to meters (a meter is a little more than a yard) and changing 1 hour to 3600 seconds means the average acceleration is about 7.66 meters/(second)^2 or 0.78g. The Corvette mass is about 1500 kilograms.

How much force (on average) is needed to produce this impressive acceleration? You could ask an expert. You could guess. You could just say it is a really big force. But, physics frees a person from uncertainty and a dependence on experts. Just believe in Newton's equation. Multiply the 1500-kilogram mass by the 7.66 meters/(second)^2 acceleration and discover that an average force of 11,490 newtons is needed to accelerate one of these racy cars.

A number like 11,490 newtons is a solution to a physics problem, but it is not really physics until one understands what this number means. One way to visualize a 11,490-newton force is to compare it with the force needed to lift something. Because 9.8 newtons is needed to hold up 1 kilogram, 11,490 newtons is enough force to lift

Figure 1.5 Fast car.

1172 kilograms, or more than a ton. This is a simplification. For a race car, the net force accelerating the car is considerably larger at the beginning of its 3.5-second journey and smaller at the end. The estimates here give only the average acceleration and the average force.

Many animals can achieve an *initial* acceleration greater than a Corvette. These include a human sprinter, a racehorse, a jaguar, and a greyhound (the animals) and an ostrich. How can that be? Even the jaguar cannot outrace a car on a quarter-mile track. The distinction lies in the time interval over which acceleration can be maintained. Because speed (starting from rest) is the product of acceleration and time, the speed prize goes to the persistent accelerator. Animals cannot maintain impressive accelerations for even a second. Long before 3.5 seconds have elapsed, the animal speed has peaked while the car speed is still increasing.

Cars are usually characterized by their power rather than the force accelerating the car. Force produces the acceleration, but power is needed at high speeds because power is the product of force and speed. Power and its relation to energy are described in Chapter 2.

1.2.13 Acceleration without Speed Change: Curved Paths

When objects are not moving in a straight line, acceleration is more complicated. Speed is no longer simply the acceleration multiplied by the time. An important example is circular motion at a constant speed. There is acceleration toward the center of the circle (Figure 1.6).

This may seem confusing because the circular path never brings the object closer to the center and the speed never changes. To make sense of acceleration in this case, it is important to distinguish between *speed* and *velocity*. Velocity has a direction, but speed only tells how fast something is going. Acceleration means a change in velocity, and for circular motion, the velocity can change even when the speed is constant. The useful formula for circular motion is

$$Acceleration = \frac{(Speed)^2}{Radius}$$

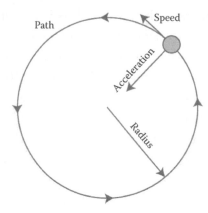

Figure 1.6 An object moving in a circle accelerates toward the center of the circle.

Deriving this formula is tedious, but it makes sense. Anyone driving a car in a circle knows faster speeds produce more sideways acceleration; larger circles reduce the acceleration. You know about the acceleration of curved paths by being thrown to the side of the car as the wheels screech on the road around the curve. For example, if a car travels at 25 meters/second (about 56 miles per hour) and travels the circle of Figure 1.6 with a radius of 50 meters (about half a football field) the acceleration toward the center of the circle would be 12.5 meters/second². This is greater than the acceleration of gravity and would be uncomfortable.

The acceleration associated with circular motion has plenty of applications in everyday life. You can swing a bucket of water over your head and not get wet if the acceleration produced by the circular motion is greater than the acceleration of gravity. When David confronted Goliath with his sling, a rock circled his head. The force needed to maintain the circular motion was a pull on the cords holding the rock. When the string was released, only the acceleration of gravity remained, and the trajectory became the ballistic path of Figure 1.7 directed toward Goliath's head.

On a much larger scale, the acceleration of circular motion means Earth does not fall into the sun and the moon will not come down. Gravity is the force that produces the acceleration needed to keep planets and moons in circular (technically elliptical) orbits.

Figure 1.7 David's rock follows a trajectory determined by its initial direction and speed as well as the acceleration of gravity.

1.3 Gravity

Newton did not reward us with just one equation. His "universal law of gravity" is the second piece to the puzzle that explains why all objects fall at the same rate. Because the law is universal, it describes much more than falling apples. Planetary orbits and the path of the moon are also explained by the universal law of gravity. However, *explain* is not quite the right word. Newton said, "I have told you how it moves, not why." No one really knows why.

The universal law of gravity states that all mass pairs (M and m) attract each other. The force of attraction is

$$F = -G\frac{Mm}{r^2}$$

The gravitational constant G (sometimes called "big G" to distinguish from Earth's gravity, "little g") is one of the fundamental constants of physics. This equation tells us that mass M is attracted toward mass m by the force F; by convention, the negative sign denotes attraction.

There is symmetry. Even though the masses may be very different, the larger mass always feels exactly the same force as the smaller mass. The equal and opposite forces are an example of Newton's third

Figure 1.8 Equal and opposite forces characterize the gravitational attraction.

law that is generally applicable—even when the force is not gravity (Figure 1.8):

> *Third law*: When two bodies exert forces on each other, these action and reaction forces are equal in magnitude but opposite in direction.

1.3.1 Gravitational Field

If a tree falls in a forest and no one is around to hear it, is there a sound? From a physics point of view, this is an easy question. Of course there is a sound because sound is a vibration of the air. A listener is not required. In the same sense, the world does not disappear when we close our eyes.

An analogous question for gravity is more interesting. Two masses are needed for the gravitational force. What if only mass M is present (Figure 1.9)? The most satisfactory answer to this question postulates a "gravitational field." Mass M produces this field without regard to the presence or absence of mass m. With this physical picture, the force on mass m is the product of m and the gravitational field produced by M.

A related question asks how gravity can propagate through empty space. In the unlikely event that a tree falls in a vacuum, there would be no sound. For many years, some believed there must be something analogous to an atmosphere transmitting the force of gravity, but

Figure 1.9 The mutual attraction of two masses. If mass m is removed, both forces disappear, but the gravitational field of mass M remains.

this is not the case. Gravity attracts in a perfect vacuum. This means "empty space" is not as empty as one might think.

1.3.2 Inverse Square Law

The squared quantity r that appears in the denominator of the universal law of gravity is the distance between the two masses. Thus, the mutual attraction decreases rapidly as this distance increases, as is shown in Figure 1.10. The inverse square relation pervades physics (e.g., electrostatics). It is a consequence of the geometry of our three-dimensional world. It is no coincidence that the surface area of a sphere of radius r is proportional to r^2, so if you multiply the sphere's surface area by the force you obtain a number that does not depend on r.

An example illustrates the inverse square relation. A painter has just 1 liter (about a quart) of paint. She is assigned the job of using all her paint to cover a sphere of radius r. How thick will the layer of paint be, and how will the thickness vary with the sphere radius? The answer is the inverse square relation; the larger the sphere is, the thinner the paint layer will be. Gravity is like this. The mass M "paints" all of space with a field that will pull on any other mass m. Just as the paint layer becomes thinner and thinner as the sphere radius increases, gravity thins at larger distances.

"Everybody" knows about the inverse square law. In *War and Peace*, Leo Tolstoy described Nicolas Rostov's approach to home. He used physics to explain why absence does *not* make the heart grow fonder:

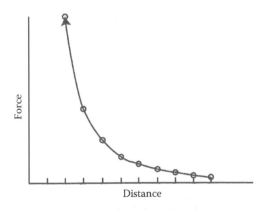

Figure 1.10   The inverse square law means the force decreases rapidly with distance.

The nearer he got, the more intense, far more intense, were his thoughts of home (as though moral feeling were subject to the law of acceleration in inverse ratio with the square of the distance). At the station nearest to Otrandnoe he gave the sledge driver a tip of three roubles, and ran breathless up the steps of his home, like a boy.

1.3.3 Measuring G

Newton could not derive the gravitational constant (big G) in his law of gravity. Even today, no one has found a theory that determines G. One must rely on experiments. The first accurate measurement of G (the Cavendish experiment) was completed more than 100 years after Newton's *Principia*. Cavendish's drawing of his experiment is shown in Figure 1.11.

In 1797–1798, Henry Cavendish suspended a rod from a thin fiber. Masses were placed on the ends of the rod and other masses were placed nearby. Gravity pulled masses toward each other. Even though the tiny force was only large enough to lift a grain of sand, it managed to twist the wire just enough to be measured. As one would expect for such a delicate measurement, many things could go wrong.

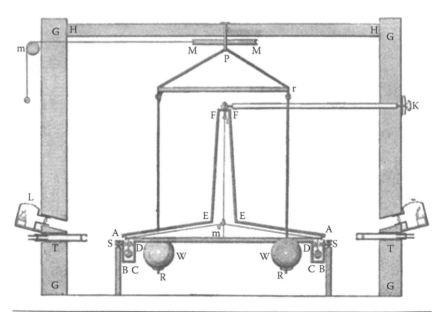

Figure 1.11 The experimental apparatus used to obtain the first accurate measure of the universal gravitational constant G.

The experiment was tedious, frustrating, and expensive. Fortunately, Henry Cavendish was exactly the right person for the job. He was meticulous, patient, very rich, and very shy.

With no significant social life and plenty of money, Henry Cavendish accomplished what few others would have been capable. His measurement of *G* was accurate to about 1%. A better value was not obtained for 97 years. Today's value is

$$G = 6.675 \times 10^{-11} \frac{\text{newton} \times (\text{meter})^2}{(\text{kilogram})^2}$$

The units assigned to *G* remind one that the force is newtons, the distance is meters, and the masses are kilograms. The 10^{-11} means divide by 10 eleven times. Big *G* stubbornly remains the fundamental constant whose value is known with the least precision (accurate today to about 1/10%).

1.3.4 Mutual Attraction

A contrived and approximate example shows that masses must be very large before gravity is significant. Assume Sonya's mass is *m* = 50 kilograms and Nicolas is bigger, with *M* = 75 kilograms. They are separated by *r* = 2 meters. Gravity attracts Sonya to Nicolas and Nicolas to Sonya. Substituting *M*, *m*, *r*, and *G* into the universal law of gravity yields a mutual attraction that is only 5/8 micronewtons for both Sonya and Nicolas (*micro* means divide by 1 million). This is less than the force needed to lift a grain of salt, so Nicolas and Sonya are unlikely to notice their mutual gravitational attraction. If Sonya and Nicolas were sitting on a frictionless surface, one could use $\vec{F} = m\vec{a}$ to compute their accelerations toward each other. Because Sonya is lighter, she would move faster, but it would take a very long time before they slid into each other's arms. If Nicolas and Sonya were separated by only 1 meter instead of 2 meters, the inverse square law says the force would be quadrupled, but it would still be insignificant.

1.3.5 Falling Objects, Earth's Mass, and Weight Loss

All objects fall at the same rate because gravity pulls harder on more massive objects, but massive objects need a larger force to accelerate. Nicolas falls off a ladder. His acceleration is $g = 9.8$ meters/(second)^2. His acceleration is also obtained by combining the two equations $F = ma$ and $F = -GMm / r^2$. Using m as Nicolas's mass, M as Earth's mass, and r as Earth's radius, and knowing that the acceleration a is the acceleration of gravity g on Earth, the result is

$$g = G \frac{Earth\ mass}{(Earth\ radius)^2}$$

The mass of Nicolas does not appear in the result. Sonya would have the same acceleration if she fell from the ladder. The two equivalent expressions for the force on Nicolas are shown in Figure 1.12.

The equation determines Earth's mass, assuming its radius, g, and G are all known. The result is $M(Earth) = 6 \times 10^{24}$ kilograms. This huge mass gives one very little insight. Dividing the mass by Earth's volume is more interesting. It tells us the density (mass/volume) of

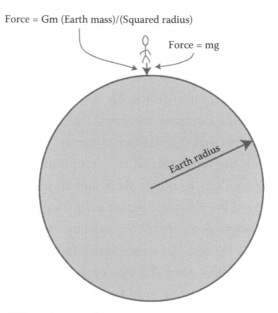

Force = Gm (Earth mass)/(Squared radius)

Force = mg

Earth radius

Figure 1.12 Two ways of computing the force on a person must give the same result. This means the mass of Earth can be determined.

Earth is about 5520 kilograms per cubic meter. This is about twice the density of granite but less than the density of iron. The density of water is 1000 kilograms per cubic meter. These densities give a hint about the composition of Earth's interior.

The following is a key to losing weight. The constant G is fundamental. The acceleration of gravity g that determines weight, is not. It varies slightly with location. Earth's rotation means a person on the equator is really moving in a circle. The speed is the Earth's circumference divided by about 24 hours. The circle radius is the Earth's radius. The formula $a = v^2 / r$ means a person sitting on the equator is accelerating toward Earth's center at about a 30th of a meter per second squared. The acceleration of gravity is decreased by this amount, but this small correction is not the whole story. Earth's rotation causes it to bulge slightly at the equator, increasing the distance to the center. The result is a decrease of g by about 0.6%. The two corrections mean a 60-kilogram person who weighs 132 pounds at the North Pole will lose 0.8 pound by moving to the equator. You can also lose weight (but not mass) by climbing a mountain because you will be further from Earth's center.

1.4 Solar System and Beyond

How high the moon? This good question has intrigued poets and scientists for a long time. It is one of many questions one can ask about our solar system. Some answers are listed in Table 1.1.

Today, spacecraft, lasers, and fast computers mean all the quantities in Table 1.1 (and much more) can be determined with great precision. The descriptions that follow do not praise the modern technology that accurately describes our solar system. Rather, they remind us that past scientists were clever. Placed on an isolated island, how many of us could come up with the common-sense observations and careful logic they used to understand the skies?

Table 1.1 Some Characteristics of Our Solar System

	EARTH	**MOON**	**SUN**
Radius	6370 kilometers	¼ Earth's radius	110 Earth radii
Mass	5.97×10^{24} kilograms	1/(80) Earth's mass	333,000 Earth masses
Distance from Earth		60 Earth radii	390 Earth–moon distances

1.4.1 Distance-to-Radius Ratios

Start with the easiest observation: The ratio [distance to moon] divided by [moon radius] can be measured with the spare change in your pocket and a meter stick. A penny held 2 meters from your eye is just big enough to cover the full moon, as shown in Figure 1.13. Because they appear to be the same size, the distance-to-radius ratios must be the same for the penny and the moon. Two meters is a little more than 200 times the penny radius. From this, one knows that the distance to the moon is a little more than 200 times the moon's radius. (The actual ratio for the moon is about 220, and the moon's elliptical orbit means the Earth-moon distance varies by about 10% during 1 month.) The moon appears to be bigger when it is near the horizon instead of overhead. The penny and meter stick measurement proves that this apparent variation in moon size is an optical illusion.

One could perform the same experiment to measure the ratio [distance to sun] divided by [sun radius], but it is neither necessary nor advisable to stare at the blinding sun. The amazing coincidence of the nearly identical apparent sizes of the sun and the moon are

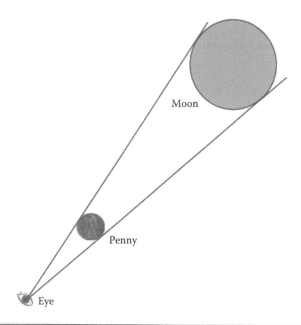

Figure 1.13 Because the penny and the moon appear to be the same size, the distance-to-radius ratios are the same for the two objects. This example presents an implausibly close moon. The actual moon distance needs additional observations.

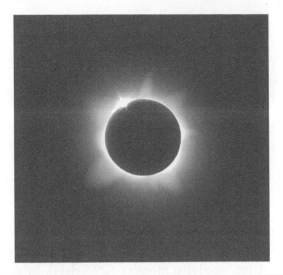

Figure 1.14 A solar eclipse shows that the moon just barely obscures the sun.

dramatically seen in a rare solar eclipse of Figure 1.14. Of course, the sun is really much larger because it is much farther away.

1.4.2 Distance to the Moon

The distance to the moon (rather than the distance-to-radius ratio) is more difficult to measure. It is easy to imagine that the moon is very close or very far. In *The Song of Hiawatha*, Henry Wadsworth Longfellow tells us that Hiawatha's grandmother "daughter of the moon Nokomis" fell from the moon. This suggests a pretty close moon. Other views of the universe placed the moon on a "celestial sphere" a great distance from Earth. Neither poetic description of the Earth-moon distance is quantitative.

Starting around 300 BC, the Greeks invented some tricks that gave reasonable estimates of the distance to the moon. One of these clever tricks is described here.

A comparison of the moon's distance to Earth's diameter can be obtained from a lunar eclipse. A slowly moving moon will remain in Earth's shadow for a relatively long time. A speedy moon will pass through the shadow much more quickly. Because a lunar eclipse lasts about 3 hours, we know the moon's speed is two Earth radii (Earth's diameter) divided by the 3-hour duration of the eclipse. Another

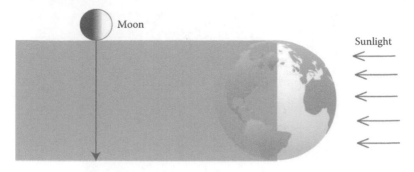

Figure 1.15 The time it takes the moon to pass through Earth's shadow determines its speed, provided the size of Earth is known.

measure of the same speed is the distance traveled by the moon as it orbits Earth (the circumference or $2\pi \times$ [distance to moon]) divided by the time (one lunar month or 29.5 days). In about 270 BC, Aristarchus of Samos equated these two expressions for the lunar speed and determined the distance to the moon to be between 62 and 73 times Earth's radius. This is reasonably close to the correct ratio of 60. Because the sun is large, Earth's shadow is fuzzy at the edges, and the duration of the eclipse is not as clear as Figure 1.15 suggests, so Aristarchus was probably lucky to obtain such an accurate answer.

Aristarchus had a view of our solar system that was far ahead of its time. He was the first to say the planets circle the sun. Unfortunately, the Greek establishment (notably Aristotle) did not accept his ideas. Nicolaus Copernicus revived the heliocentric theory 1800 years later (in 1543). Copernicus died on nearly the same day his book was published, so he escaped the political difficulties faced by Galileo, who believed Copernicus was correct.

1.4.3 Size of Earth

About 70 years after Aristarchus determined that the moon was 62 to 73 Earth radii away, Eratosthenes completed the moon-distance determination by estimating the size of Earth. The idea was to compare the direction of the sun's rays at two distant points, as shown in Figure 1.16. On a day when the noonday sun was directly overhead in Syene (Aswan in Egypt), the sun appeared to be 1/50 of a circle south of overhead in Alexandria. Because Alexandria is about 900 kilometers north of Syene and 1/50 of a circle away, 50 of these

Sunlight

Figure 1.16 "Straight up" is parallel to the sun's rays at one point, but it is tipped at an angle for a point further north. The angle and the distance between the points determine the size of Earth.

900-kilometer trips would take one all the way around Earth. Thus, Earth's circumference is about 50×900 or 45,000 kilometers. The corresponding Earth radius, a little more than 7000 kilometers, is a reasonable rough estimate. The 10% overestimate can be forgiven because it was not easy to measure the distance between cities in ancient times. Also, distance scales were not standardized 2200 years ago. Uncertainty of the length of a "cubit," means it is hard to judge the accuracy of this early estimate of Earth's size.

1.4.4 Size of the Sun and Distance to the Sun

The size of the sun and Earth's distance to the sun could not be obtained by the ancient Greeks. The first fairly accurate results were obtained much later from observations of Venus passing in front of the sun (transit of Venus). Results from the 1761 and especially the 1769 transit yielded results that were accurate to better than 3%. These measurements were obtained at great personal cost to some of the scientists, who went to the ends of the Earth to see how the sun position varied with respect to the Venus pivot point. A more accurate solar distance was not obtained for 100 years.

As a first step to the solar distance, one uses the geometry of Figure 1.17 to show that Venus is about 70% as far from the sun as Earth. Then, if one draws lines from two different points on Earth that pass through Venus during the transit, these lines will hit two different points on the sun. The geometry of Figure 1.18 shows that the separation of the points on the sun will be 7/3 the separation of the points on Earth. This is the key to estimating the size of the sun,

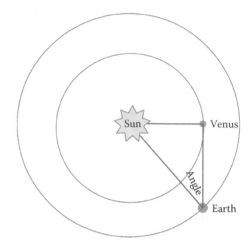

Figure 1.17 The largest angle between Venus and the sun determines the ratio of the Venus–sun distance to the Earth–sun distance.

Figure 1.18 Sighting past Venus from two points on Earth compares the sun's size to Earth's size.

which in turn can give the distance to the sun (because the distance is about 220 times the radius).

The sun is very far away. Earth must move at high speed to circle the sun in a year. Dividing the circumference of our orbit by 1 year shows that we are moving at about 85 times the speed of sound. Because everything on Earth is moving at the same speed, this motion is hardly noticed. By comparison, the additional speed caused by Earth's rotation is relatively small, being only about 1.3 times the speed of sound. All these speeds are tiny compared to the speed of light, which is close to 1 million times the speed of sound.

1.4.5 Kepler's Laws

Johannes Kepler formulated his three laws in the early 1600s. They were based on a careful analysis of Tycho Brahe's detailed astronomy measurements. Kepler was a contemporary of Galileo (about 8 years younger). His three laws, stated in modern language are as follows:

1. The orbit of every planet is an ellipse with the sun at one of the foci.
2. A line joining a planet and the sun sweeps out equal areas in equal time intervals.
3. The square of the time to go around the sun is proportional to the cube of the sum of the shortest and largest distances from the sun.

Kepler's laws can be derived from Newtonian mechanics, but only with considerable effort. Kepler's laws represent a big step in science, but he was not on the verge of understanding gravity. He believed the sun's rotation dragged the planets along their orbital paths.

1.4.5.1 First Law   Even with modern mathematics, a derivation of Kepler's first law (elliptical orbits) from Newton's equation and the universal law of gravity requires some imaginative tricks. The trickiness makes one wonder if Newton really could prove Kepler's first law. Halley (of Halley's comet) reportedly asked Newton about orbit shapes, and Newton is supposed to have said the answer was clearly an ellipse. Just because someone says something is obvious does not constitute a proof, even if the person asserting the "obvious" is Isaac Newton. Richard Feynman (another famous scientist) attempted to derive the elliptical orbits using the mathematics that Newton would have known. Feynman resorted to some of his own ideas. Apparently, even Feynman had trouble following Newton. The resurrected work *Feynman's Lost Lecture: The Motion of Planets around the Sun* is not an easy read, but it is far easier to read than Newton's *Principia*.

1.4.5.2 Second Law   Kepler's second law (equal areas in equal times) follows from the principle of conserved angular momentum. Angular momentum and a demonstration of the second law are provided in Section 2.3.

1.4.5.3 Third Law   One can deduce a lot about our solar system from Kepler's third law and its generalization. Because Kepler died 12 years before Newton was born, he had no understanding of the underlying physics. Today, we can derive Kepler's third law, and the derivation is not a "killer" when orbits are approximated by circles instead of ellipses. One needs three formulas and some algebra. The formulas are Newton's equation ($\vec{F} = m\vec{a}$), the universal law of gravity ($F = GMm / r^2$), and the expression for the acceleration of a circular path, $a = v^2 / r$. Then, algebra gives "Kepler's third law formula":

$$(Period)^2 = \frac{(2\pi)^2}{G \times (central\ mass)} \times (Radius)^3$$

Thanks to Newton, this formula is more than Kepler's third law because the constant multiplying by the cubed radius is determined by the gravitational constant G and the central mass. For planets orbiting the sun, the central mass is the solar mass.

1.4.6 Applications of Kepler's Third Law Formula

1.4.6.1 Mass of the Sun This first application considers Earth's orbit about the sun. For this case, "Period" is the one year needed to circle the sun. "Radius" is the distance between Earth and the sun. This distance was not well known until the "transit of Venus" observations were done 130 years after Kepler's death. The gravitational constant was determined by Cavendish in 1798. By 1800, only one quantity in the Kepler third law formula was not known: That is the "central mass" holding Earth in its orbit, and this is the sun's mass. When everything is known except for one quantity, a formula supplies the answer for the last piece. The result is impressive. The sun is about a third of a million times as massive as Earth. This huge mass means nearly all (except about seven parts per thousand) of our solar system mass lies in the sun.

1.4.6.2 Distance Comparisons The Kepler third law formula allows one to determine relative distances from the sun if one knows the time it takes to orbit the sun. The example of Figure 1.19 shows how this works.

A year on Mars is nearly 8 Mercury years. The square of 8 is 64. The cube of 4 is 64. Thus, the formula tells us that Mars is nearly four times further from the sun as Mercury. This comparison does not need information about the gravitational constant or the mass of the sun.

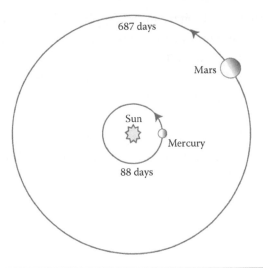

Figure 1.19 Because the Mars year is eight times the Mercury year, Mars is four times as far from the sun.

Actually, this example is historically backward. Kepler deduced his laws in part from measurements of planet periods and orbit radii.

1.4.6.3 Lunar Distance For the moon circling Earth, the Period is 1 month; the Radius is the distance to the moon, and the Earth mass is the "central mass." The Earth mass is known (again thanks to Cavendish). That means one can solve for the Radius. It is a credit to ancient science that this more precise method gives a result close to the 2000-year-old Greek measurement based on the lunar eclipse. Today, laser reflections off the moon give its distance with extraordinary accuracy. The reflections also show that the moon moves about 3.82 centimeters further from Earth each year. Because the moon is moving away, the Kepler third law formula tells us that next year it will take the moon an extra 0.38 milliseconds to orbit Earth. The lunar months are getting longer.

1.4.7 Jupiter Mass

Jupiter has 67 known moons and probably more. The largest is Ganymede. It was discovered by Galileo. (Galileo's moon observations got him into a lot of trouble with the Catholic Church.) Ganymede's period and its distance from Jupiter can be observed.

The only unknown is Jupiter's mass. Applying Kepler's third law formula shows that Jupiter's mass is about 320 times the mass of Earth. Jupiter's volume is about 1300 Earth's mass, so its density is less than a quarter of Earth's density, only slightly denser than water. Jupiter's density is low because it is mostly hydrogen. Near its core, the very high pressure makes hydrogen metallic. The electrical conductivity of this metal is probably the source of Jupiter's large magnetic field. The same calculation method that determines the mass of Jupiter can give the mass of any planet with a moon (Mars, Venus, Saturn, Uranus, and Neptune, but not moonless Mercury).

1.4.8 Dark Matter

"Dark matter" is not that new. Fritz Zwicky coined the term in 1933 while considering the motion of galactic clusters. Later observations of galaxy rotation, pioneered by Vera Rubin starting in the 1950s, and a variety of more recent astronomical observations and calculations have made it clear that dark matter pervades the universe on all distance scales.

The clearest evidence for dark matter comes from galaxy rotation rates. Just as moons orbit planets and planets orbit stars, stars orbit their galactic centers. Modern astronomy gives the distances to galaxies and the rate at which stars are circling the galaxies. Thus, a star's period and radius with respect to the galactic center are approximately known. Kepler's third law formula then yields the mass of the galaxy. This time there is a surprise. By counting the stars and estimating their average mass, one can obtain an alternative value for the galactic mass. The two results do not come close to agreeing. The orbiting stars appear to be moving much too quickly. There is only one sensible conclusion consistent with all observations:

There is invisible and undetected material attracting the stars to the galactic center and shortening the period.

Although speculations abound, almost nothing yet is known about dark matter besides its gravity (Figure 1.20). I find it appealing that an application of 300-year-old physics contributed to the twentieth-century discovery of dark matter.

Figure 1.20 The best-available picture of dark matter.

1.4.9 Moon Mass

Nothing orbits the moon, so the Kepler third law formula cannot determine the lunar mass. The moon's period of 29 days (a month) would be the same if it were made of lead or cheese. It is strange but true that the moon's mass is more difficult to estimate than the mass of distant planets.

Newton devised an alternate way to calculate the moon's mass using tides. Both the moon and the sun contribute to tides on Earth. Newton knew that spring tides (sun-Earth-moon in a line) are significantly higher than neap tides (sun-Earth-moon makes a right angle). These tides differ because both the sun and moon pull the water on the near side of Earth and leave the water on the far side behind. By comparing spring tides and neap tides, Newton could compare the forces from the sun and the moon. From this, he obtained an estimate of the lunar mass. In principle, Newton's idea was clever. In practice, the complexity of ocean motion caused Newton to overestimate the moon mass by about a factor of three.

The most surprising consequence of Newton's overestimate of the moon mass was Edmond Halley's unquestioning acceptance of the result. Because Newton claimed that Earth was only about 5/9 as dense as the moon, Edmond (of Halley's comet) guessed that Earth was hollow with 4/9 empty space. He published a paper in 1692 with a catchy title:

"An account of the cause of the change of the variation of the magnetical needle with an hypothesis of the structure of the internal parts of the earth: as it was proposed to the Royal Society in one of their later meetings"

and it included Figure 1.21.

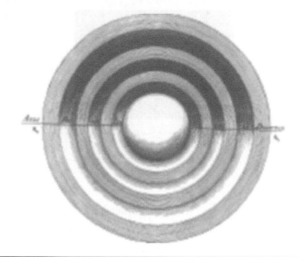

Figure 1.21 Halley's model of a hollow Earth.

As the title suggests, Halley claimed his hollow-Earth model also explained some of Earth's magnetic properties and speculated that gas escaping from the center regions produced the aurora. The possibility of a hollow Earth, perhaps with an interesting "underground," has a long tradition, appearing in many cultures and religions from ancient to recent times.

If tides are too complicated, how does one estimate the moon's mass? Today, with artificial satellites flying around the moon, the solution to the problem is just another application of Kepler's third law formula.

1.4.10 Extrasolar Planets

Our sun has eight planets. There are roughly 200 billion stars in our galaxy, the Milky Way. It seems sensible that many of these stars should also have planets, and some of the planets could be similar to Earth. In recent years, clear evidence for planets orbiting other stars has been mounting. The number of observed "exoplanets" is increasing rapidly. Some of the exoplanet evidence is seen as a decrease in star brightness when the planet passes in front of the star. One is bound to wonder what, if anything, inhabits these planets.

1.5 Causality and Chaos

1.5.1 Gravity Is Not the Only Force

If gravity were the only force, our world would be very different. The force of gravity always pulls things together, so it would be impossible for anything to push on anything else. Objects would have no size because there would be nothing to keep two things from being at the same place at the same time. Something is missing. Actually, a lot is missing. Additional repulsive forces and quantum mechanics give objects shape and size and hardness.

One of the other forces is electrostatic repulsion and attraction. It is described by a formula similar to the universal law of gravity. Other forces, like those that hold the nucleus together, are more complicated. But, given these forces, Newton's equation, $F = ma$, and the law of equal and opposite forces remain valid. And, of course, the laws generalize to more than two objects. This leads to philosophical discussions of determinism. To see the classical reasoning behind the causality dilemma, assume all forces are known and accept the idea that they explain solids, liquids, gases, pressure, collisions, and everything else.

1.5.2 Causality

Newton's equation tells us that forces determine the acceleration of all objects, whether these forces are gravity, electromagnetism, or something else. Regardless of the forces, acceleration determines future positions. In principle, this applies not only to simple problems like the orbit of the moon but also to the motion of any collection of objects. Thus, the future of anything is predicable using the laws of physics. This leads to some controversial consequences, as eloquently stated by Pierre Simon Laplace in 1814:

> We may regard the present state of the universe as the effect of its past and the cause of its future. An intellect which at a certain moment would know all forces that set nature in motion, and all positions of all items of which nature is composed, if this intellect were also vast enough to submit these data to analysis, it would embrace in a single formula the movements of the greatest bodies of the universe and those

of the tiniest atom; for such an intellect nothing would be uncertain and the future just like the past would be present before its eyes.

Even though there is no mythical intellect, later called the "Laplace demon," its feasibility is disturbing. If the future can be predicted, then one has trouble finding room for free will or divine intervention. On this question, Napoleon Bonaparte and Laplace were reported to have engaged in the following conversation:

Napoleon: How can this be! You made the system of the world, you explain the laws of all creation, but in all your books you speak not once of the existence of God!
Laplace: I had no need of that hypothesis.

At the time of Laplace, people believed that Newton's laws really were "laws," with no exceptions allowed. Although modern physics goes far beyond Newton, philosophical questions about determinism and causality remain and are particularly perplexing for black hole physics (see Section 8.3.1). These questions will not be answered here. Perhaps they have no answers.

1.5.3 Chaos = Unpredictable

Consider a more practical question. Assuming the validity of the Newtonian approach, is it possible that a great intellect (or a giant computer) really could predict the future? In many cases, the answer is "no" because even microscopic changes today can grow explosively, leading to big changes tomorrow. No computer can be accurate enough to deal with this growth.

As an example, balance a pencil on its end. If it falls to the left, you become (or remain) a vegetarian. A fall to the right leads to the opposite dietary choice. The direction of the pencil's fall will have an enormous effect on some future chicken. The initial orientation of the pencil determines the direction it falls, but one could use the most precise equipment possible to make the pencil vertical. Then, you would not know the direction it would fall—but it will eventually fall. An error too small to measure becomes consequential.

Important early studies of a much more complicated version of unpredictability were done in 1887 by the extraordinarily talented

Henri Poincare. The results of Poincare and others suggested that one need not perfectly balance a pencil to be confronted with unpredictability. Analogies to delicately balanced pencils are hiding everywhere. However, it took some time to realize the broad implications of this. In the early days of computers, many people, including the mathematical genius John von Neumann, believed weather predictions extending out a month or more would be possible as computers improved. It is painfully clear that this prediction of predictions has not come true.

A famous speculation on how tiny initial effects could affect the future was put in the form of a question in a 1972 talk by Edward Lorenz:

Does the flap of a butterfly's wings in Brazil set off a tornado in Texas?

Lorenz had studied mathematical simulations of weather since the 1960s. His work suggested that a Brazilian butterfly really could pose a problem for Texas. Although mathematicians already knew of many formal examples for which small initial changes produce large and unpredictable final outcomes, they failed to popularize their knowledge with anything as lovely as the butterfly–tornado pair of Figure 1.22.

Sometimes, chaotic behavior can mask itself for a long time. A system can appear stable and predictable until—suddenly everything changes. The unpredictable reversal of Earth's magnetic poles that typically occurs only after thousands of years *might* be an example of this type of chaos. A simplified graph of Earth's magnetic history is shown in Figure 1.23.

Figure 1.22   A butterfly leads to a tornado.

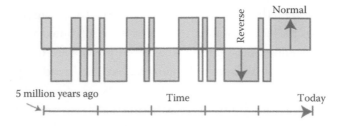

Figure 1.23 A simplified picture of Earth's magnetic field over the ages. No one can predict the next magnetic field reversal.

The determination of the magnetic field history is a good detective story. When hot rocks with some iron compounds cool, they acquire a weak permanent magnetization in the direction of Earth's field at the time of cooling. Finding old, but undisturbed, rocks with reversed magnetism indicates a reversal in Earth's field. The age of the rocks when they cooled must also be determined. Sometimes, this is done by analyzing the radioactivity in the rocks.

These examples do not mean everything is unpredictable. We know the sun will rise tomorrow, and we can predict lunar eclipses years ahead of time. An example of both the predictable and the unpredictable is seen in Saturn's rings (Figure 1.24). The rings are made of a large number of fairly small objects (1 centimeter to a few meters) circling the planet in the plane of its equator at distances from around 7000 kilometers to 80,000 kilometers. There are many curious gaps in the rings at radii where few objects are orbiting the planet. They are

Figure 1.24 Saturn and its rings.

missing because the orbits at special radii are unstable. Their motion cannot be predicted any better than a toss of a coin or the fall of a vertical pencil. At other radii, the motion is nearly as predictable as our moon's orbit. Some of the orbits are unpredictable because Saturn's moons deflect their paths. After a long time, a moon can "kick" objects out of their unstable orbits. Recent satellite observations of Saturn's rings showed an exceedingly complicated geometry, with many minigaps and other structures.

Saturn's rings are an example for which formal mathematics may give some insight. A result called the KAM theorem predicts that some systems should exhibit complicated motion, some parts of which should be predictable and other parts not. The predictable and unpredictable behaviors are delicately interlaced. The original idea was from Andrey Kolmogorov in 1954 (the K in KAM). Rigorous proofs and extensions by Vladimir Arnold and Jürgen Moser in the 1960s added the A and the M.

It is natural to wonder why Saturn's rings are so complicated. In comparison, the moon's orbit around Earth seems relatively straight-forward. The hidden trick lies in the simplicity of two-body problems. As long as one considers only Earth and the moon, the orbit is simple. But, if one desired an exactly correct solution, one would have to consider the influence of the sun on the moon's orbit. This is a three-body problem. Mechanics problems involving more than two objects present truly annoying difficulties that sometimes result in the unpredictability one sees in a coin toss or Saturn's rings.

An exact description of our solar system would require the solution of a complicated many-body problem. For example, Jupiter has a measureable influence on Earth's orbit, and there is a proof that the solar system is not stable. Of course, this does not mean the sun may not rise tomorrow. The instability means that we cannot be assured of maintaining our pleasant distance from the sun forever, but forever is a very long time.

Analysis of chaotic systems is beset with complexity. Applications of astoundingly difficult mathematics have their limitations, and physical intuition frequently fails. However, the combinations of formal math, extensive numerical work, and observations have led to growing insight into the nature of chaos.

2

MOMENTUM, ANGULAR MOMENTUM, AND ENERGY

2.1 Introduction

When a system is isolated from the outside world, its momentum, angular momentum, and energy are all conserved. *Conserved* means they do not change. The constant value of these quantities allows one to understand a wide range of phenomena. For mechanical systems, the conservation laws for all three follow from Newton's laws, but their conservation extends beyond Newtonian mechanics to quantum mechanics and relativity. In this more general context, conservation laws follow from basic symmetries of space and time. The results of a physics experiment (isolated from the outside world) are unchanged if the equipment is moved to a new place, aimed in a different direction or done on a different day. These uniformities of space-time yield the conservation laws for momentum, angular momentum, and energy. Other symmetries, such as time reversal and reflection invariance, are associated with additional conservation laws.

Conservation laws simplify the descriptions of pole vaults, tides, billiards, clocks, rockets, gyroscopes, collision, violins, and much more. Examples like these illustrate the central role of momentum, angular momentum, and energy.

2.2 Momentum

The momentum $\vec{p}$ of an object with mass m moving with a velocity $\vec{v}$ is the product of its mass and velocity:

$$\vec{p} = m\vec{v}$$

One could ask why one assigns a special letter p to momentum because it is almost as easy to write mv. A justification lies in the generality of momentum that extends beyond its basic mechanical definition. Light with no mass has momentum, and particles moving close to the speed of light have an altered formula for their momentum.

2.2.1 Momentum Conservation

The total momentum of any system, no matter how complicated, is a constant if the system is not subjected to any external forces.

A key part of this law is the *if*. One must make sure there are no outside forces before one knows momentum is conserved.

For a single object, momentum conservation is an obvious consequence of Newton's equation. No force means no acceleration. No acceleration means a constant velocity and thus a constant momentum.

When two objects interact, the momentum of each object can change. However, the sum of the two momenta remains constant. The momentum gained by one is lost by the other because Newton's third law says the forces are equal and opposite. The opposite forces produce opposite momentum changes. Alternatively, viewing the two pieces as part of a larger system, the external force is zero, so the total momentum is conserved.

2.2.2 Examples

Example 2.1: Falling Objects

A man falls off a cliff. As he is accelerated toward the ground, his momentum is not conserved. It is increasing in the downward direction because he is a victim of gravity. But, just as Earth is pulling the man down, the man is pulling Earth up. Earth's momentum is increasing in the opposite direction to conserve the total momentum. Of course, no one notices massive Earth's tiny acceleration.

Example 2.2: Sonya's Mass

Nicolas is curious about Sonya's mass, but a simple question would be indiscreet. Sonya and Nicolas go ice skating, and

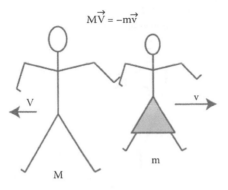

Figure 2.1 Nicolas pushes Sonya. If Sonya's mass is smaller, conservation of momentum means she will move faster.

Nicolas pushes Sonya. The partners slide away from each other on the frictionless ice surface. Conservation of momentum means Sonya will travel slower than Nicolas if her mass is larger. Instead, Nicolas discovers that Sonya's speed is one and a half times his speed. Equating the two opposing momenta allows Nicolas to determine that Sonya's mass is only two-thirds of his mass (Figure 2.1).

2.2.2.1 Recoil Recoil can be a serious problem for large guns (Figure 2.2). Early cannons were mounted on wheels so the cannon and its mount could absorb the recoil rather than tipping over. Ropes limited the recoil distance. More recent large artillery allow the barrel to compress a spring. This smoothes out the recoil, but momentum is still conserved.

2.2.2.2 Rocket Sending a rocket into space is analogous to repeated cannon fires. Thanks to conservation of momentum, the continuous

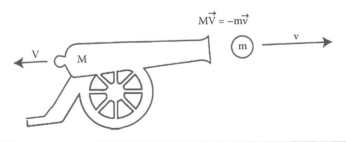

Figure 2.2 A cannon on wheels. It rolls backward after firing.

Rocket
Momentum

Gas
Momentum

Figure 2.3 The momentum increase of the expelled gas is equal to (but opposite in direction) the momentum gain of the rocket.

expelling of high-velocity gas propels the rocket (Figure 2.3). This is a tricky case to analyze because the mass of the rocket decreases as it uses its fuel. Also, part of the imparted momentum is used to counter the force of gravity.

The rocket exhaust need not push on anything to accelerate the rocket. For space travel, it is fortunate that an atmosphere is unnecessary. This idea is not intuitive, as evidenced from a cynical *New York Times* article of January 13, 1920 (Topics of the Times) that claimed rockets would not work in space.

> That Professor Goddard, with his "chair" in Clark College, and the countenancing of the Smithsonian institution, does not know the relation of action to reaction and the need to have something better than a vacuum against which to react—to say that would be absurd. Of course he only seems to lack the knowledge ladled out daily in high schools.

Forty-nine years later (July 17, 1969), on the day after the launch of Apollo 11 that landed the first human on the moon, the *Times* published the following retraction:

> It is now definitely established that a rocket can function in a vacuum. The *Times* regrets the error.

2.2.2.3 Kennedy Assassination   Luis Alvarez (a Nobel Prize winner and coinventor of the theory of dinosaur extinction by asteroid impact) used momentum conservation in an analysis of the Zapruder recording of President Kennedy's assassination. Many suspected a conspiracy because the pictures showed the president's head recoiling toward the position of the supposed lone gunman. Alvarez showed that this unexpected effect was consistent with physics.

> It will turn out that the jet [of brain matter] can carry forward more momentum than was brought by the bullet, and the head recoils backward, as a rocket recoils when its jet fuel is ejected.

The conclusion that ejected brain matter could cause a head to move in the opposite direction of the oncoming bullet is not well known. Of course, an explanation of the head motion does not disprove all conspiracy theories. Surveys consistently show that more than half of all Americans believe Lee Harvey Oswald did not act alone.

2.2.3 Drag and Lift

Bicycles, cars, and baseballs naturally slow and stop. This is not a failure of momentum conservation; it is a reminder that momentum is conserved only if there are no external forces. These forces are drag and lift.

For fluids, the frictional force slowing an object is called *drag*. It is produced by collisions with the fluid molecules. Fluid drag is a little like the slowing of a bowling ball as it crashes into an array of bowling pins, with the pins playing the role of molecules. An approach pioneered by Newton (who else?) uses conservation of momentum to estimate fluid drag.

One can approach Newton's drag formula through some qualitative observations. It makes sense that the drag should be

1. Proportional to the area A of the object because larger size would push on more fluid.
2. Proportional to the density of the fluid ρ because it takes more force to move more mass.

3. Proportional to the square of the speed: one v for distance traveled through the fluid in 1 second and another v for the speed of the collisions.

Combining these gives

$$F \approx \rho A v^2$$

Here, the $\approx$ is used instead of an equal sign because the result is only an estimate. Many experiments have checked this drag formula. It typically overestimates the drag force by about a factor of 4 for spheres moving though a fluid. One should not be surprised by the overestimate because most molecules flow around the object rather than crash into it.

2.2.3.1 Bicycle A person rides a bicycle at $v = 10$ meters/second. The effective area of the bike plus rider is $A = 1/2$ square meter. The density of air is about $\rho = 1.2$ kilograms per cubic meter. Doing the multiplication and assuming the drag formula is too large by a factor of 4 gives a modest force: $F = 15$ newtons. However, if the bicyclist is faced with a 10-meter/second headwind, the effective velocity is doubled. This quadruples the force to 60 newtons. That makes riding the bicycle a much more strenuous exercise.

2.2.3.2 Tornadoes Wind speeds in a category 3 tornado are roughly double the wind speed of a category 1 tornado. The drag formula with a squared velocity tells us the destructive forces should be four times as large for the category 3 tornado.

2.2.3.3 Water The density of water is about 800 times the density of air, so if one tries to drive a golf ball through water, the drag will be magnified by about a factor of 800. This leads to a limited range. When Alan Shepard golfed on the moon, the missing atmosphere meant the drag force was zero. Nothing slowed the golf ball on its ballistic trajectory.

2.2.3.4 Da Vinci Parachute In 1483, about 200 years before Newton's *Principia*, Leonardo da Vinci sketched a design for a parachute resembling that in Figure 2.4. Da Vinci correctly identified drag, and his cup-shaped parachute design maximized drag by catching the air. He said:

Figure 2.4   A parachute design modeled after an idea sketched by Leonardo da Vinci more than six centuries ago.

If a man is provided with a length of gummed linen cloth with a length of 12 yards on each side and 12 yards high, he can jump from any great height whatsoever without injury.

A replica using materials of da Vinci's time was tested in the year 2000, and it was found to work. However, the term *hit the ground running* really applies to da Vinci's parachute. Because the heavy cloth and supporting wooden structure weighed more than a man, artful dodging is needed to avoid being crushed by one's own parachute on landing. In the test, da Vinci's parachute was abandoned for a modern chute at the last minute to avoid this problem.

2.2.3.5 Streamlined Objects   In many situations, drag is unwanted. It slows automobiles and decreases their efficiency. The same is true for swimmers, boats, and fish. For these situations, the parachute geometry is definitely to be avoided. A "streamlined" shape like that

Figure 2.5   Drag minimization is the reason a swordfish is not shaped like a parachute.

of the swordfish in Figure 2.5 can decrease its drag by a factor of 10 or more. Fish that need to swim fast for survival have evolved into some of the most streamlined shapes. It is not a surprise that objects designed for minimum drag often look aquatic.

2.2.3.6 Lift   The drag force slows an object, but lift is a force perpendicular to the direction of motion. Lift is essential for airplanes and birds. Again, it is not a surprise that airplane wings resemble bird wings. As with drag, the lift force is proportional to the product of an area, the fluid density, and the square of the speed. The speed dependence of lift is especially noticeable for airplanes. Long acceleration on the runway is needed to obtain sufficient speed to lift the plane. The accomplishment of the Wright brothers, who are usually credited for developing the first practical airplane, is particularly impressive. They did not have access to light and powerful engines. The resulting slow plane speed meant it was a real struggle to achieve sufficient lift. The average speed of the first flight was only about 15 kilometers per hour, so a person could outrun the plane. Their specially made 9000-watt (only 12-horsepower) engine had a mass of 82 kilograms. Today, engines of this mass can easily generate 10 times as much power as the engine in the first flight (Figure 2.6).

Lift has many other applications. Windmills rotate because of the lift force on the blades. This application of lift is important for the transformation of wind into electrical power. Airplane propellers use lift with a moving propeller rather than moving air. Most of our electricity is generated by the flow of steam against windmill-like blades. These blades are shaped roughly like the rotors on turbojet engines.

Figure 2.6   The Wright brothers' first flight.

2.3 Angular Momentum

What makes the world go round? The answer is conservation of angular momentum. Why did it start rotating in the first place? Again, the answer is conservation of angular momentum. Angular momentum is the rotational analogue of ordinary momentum. Torque is the rotational analogue of force. Torque is considered first.

2.3.1 Torque

"Give me a lever long enough and a fulcrum on which to place it and I shall move the world." Archimedes may or may not have said something like this. The Earth-moving claim is so silly that one wonders: Was Archimedes prone to exaggeration just to illustrate the point shown in Figure 2.7? Exaggerated statement or not, Archimedes was the first person to give a scientific explanation of a lever. Of course, people knew how to use levers long before Archimedes showed how they worked. The next example (illustrated in Figure 2.8) is a special case that simplifies the Archimedes explanation.

A horizontal bar one meter long is held up at its ends. The mass of the bar is negligible, but a 1-kilogram mass is placed at its center. Gravity pulls the kilogram down with a force of 9.8 newtons. To keep the bar from falling, an equal upward force must be applied at the ends. Symmetry means the forces are equal, so 4.9 newtons are needed at each end. The left-hand end of the bar is a fixed pivot point.

Figure 2.7 Archimedes moving the world. (From *Mechanics Magazine*, 1824.)

The other end can move up and down, so the bar and the mass can be moved with only half the force of gravity. But, there is no free lunch. The free end must be lifted twice as far as the mass moves.

One can interchange the applied force and the mass. Placing the mass at the end and applying the force at the center would require double 9.8 newtons, but the mass movement would double the movement of the applied force. Examples of machines that apply the lever principle include the wheelbarrow (for extra force) and the bicycle (for extra distance).

A more official-sounding analysis of this lever example defines the *torque* as the product of the force and the distance from the pivot point. A torque produces angular acceleration in analogy to the ordinary acceleration produced by a force.

The torque produced by the mass in Figure 2.8 is the 9.8-newton gravity force multiplied by the half meter from the pivot: 4.9 newton-meters. The 4.9-newton force applied to the end of the bar multiplied

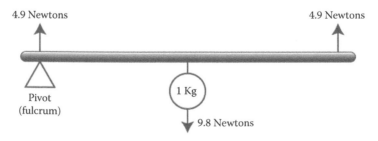

Figure 2.8 Lever. The centered 1-kilogram mass can be lifted with half the force using a lever.

Figure 2.9 A wrench provides far greater torque than a bare-handed twist of the nut.

by the 1-meter distance gives the equal and opposite 4.9-newton-meters torque. The equal and oppositely directed torques mean the bar can remain stationary.

Levers and torques appear everywhere and in many different forms. One applies a torque when opening a jar or tightening a nut, as shown in Figure 2.9. If the nut will not budge, one can use a longer wrench because torque is the product of force and distance. (In the United States, a torque is sometimes measured in foot-pounds. One foot-pound is about 1.36 newton-meters.)

Our anatomies provide many examples of torques and levers. Figure 2.10 shows an extended arm holding up a 10-kilogram mass. The upward force on the ball needed to counter gravity is $F = mg = 98$ newtons. A distance of $D = 0.4$ meters from elbow (pivot) to hand means the torque is 39.2 newton-meters. The bicep attached at $d = 0.04$ meters from the elbow must exert a force of 980 newtons

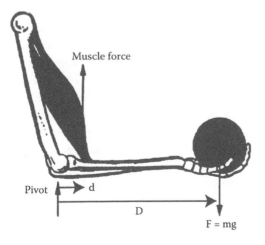

Figure 2.10 The ball held by an arm is stationary when the torque supplied by the muscle equals (with opposite sign) the torque produced by the ball.

to supply the opposing torque. However, contracting the muscle by 1 centimeter lifts the mass by 10 centimeters. The very powerful muscle forces require strong bones for support.

2.3.2 Angular Momentum Definition

Angular momentum is the product of the momentum (in the direction of rotation) and the distance from the center of rotation. Motion in and out toward the rotation center does not contribute to the angular momentum. If a system is made of several pieces, then the total angular momentum (with the same rotation axis) is the sum of the angular momenta of its pieces.

2.3.3 Angular Momentum Conservation

The total angular momentum of any system, no matter how complicated, is constant if the system is not subjected to external torques.

This law is analogous to conservation of ordinary momentum, with force replaced by torque and momentum replaced by angular momentum.

Example 2.3: Angular Momentum

Attach a rock to a string and swing it around your head. The circular motion means the rock's angular momentum is the product of its mass m, its speed v, and the length of the string l. If the string is pulled in so its length is halved, the speed v is doubled. This conserves the angular momentum mvl. The same idea applies to objects with more complicated geometries. Figure skaters rotate more rapidly when they pull in their arms. Gymnasts and divers regulate their rotation rates by compacting (faster rotation) or extending (slower rotation) their bodies. If a diver loses his or her compact position, the reduced rotation rate results in a big splash instead of a smooth entry. When you drain a bathtub, the water circulation speeds up as it moves toward the exit. Tornadoes and hurricanes can be similarly explained (in part) by conservation of angular momentum.

2.3.4 Kepler and Angular Momentum

Kepler's first law tells us that Earth is orbiting the sun in an elliptical path. Earth is 3.4% closer to the sun on January 4 than it is on July 4.

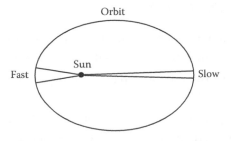

Figure 2.11   An elliptical orbit with the sun at the center. The two triangles of equal area are generated by marking the planet positions over two equal time intervals.

The inverse square law determines the sun's heating of Earth. The square of 1.034 is almost 1.07, so Earth receives 7% more solar energy in January than July. This makes the winter slightly milder north of the equator. Australians and Argentines are less fortunate. The elliptical path of Figure 2.11 exaggerates the difference between Earth's closest and furthest distance from the sun.

Kepler's second law says:

> A line joining a planet and the sun sweeps out equal areas in equal time intervals.

This law is a consequence of angular momentum conservation. There is no torque on planetary orbits because gravity pulls inward toward the sun. Thus, the angular momentum (with the sun as the pivot) applies to all the planets orbiting the sun.

The two equal areas are shown in Figure 2.11. The narrow triangle on the right, denoted "slow," is obtained from lines drawn from the sun to the planet on two sequential days. The lines are close together because the planet is moving slowly when it is furthest from the sun. The area on the left, denoted "fast," is obtained the same way. However, the lines on subsequent days are not as close because the planet is moving faster. The rapid motion when the planet is close to the sun means the triangle base is larger by exactly the amount needed to give the two triangles equal areas. Generalization of the equal-area deduction to other points on the orbit is an exercise in geometry that completes the proof of Kepler's second law.

2.3.5 Is a Day Really 24 Hours?

Conserved angular momentum keeps Earth rotating at a constant rate, so we are confident the sun will rise tomorrow. Our units of time originated in the reliable length of a day. Everyone knows that 24 hours is the time between noon and noon the next day. But, if one wants to be fussy, time is not that simple. An hour is not exactly a 24th of a day. Instead, an hour is a fixed number of oscillations of an atomic clock that pays no attention to the position of the sun.

2.3.5.1 Orbit Corrections Because we orbit the sun, a day is a little longer than the time it takes Earth to make a 360-degree rotation on its axis. This is illustrated (and exaggerated) in Figure 2.12. The hypothetical planet in this figure orbits the sun in only 20 of its days. "Noon" occurs when the arrow on this rotating planet aims directly at the sun. The orbital motion means it must rotate a little more than 360 degrees between consecutive noontimes.

Similar geometry applies to Earth, but the effect is much smaller because it takes 365 days to orbit the sun. Our unit of time is based on the noon-to-noon interval, so it takes about 3.9 minutes less than 24 hours for Earth to make one 360-degree rotation.

Seasonal variations of "one day" occur because of Earth's elliptical orbit. Figure 2.12 shows that a planet must rotate further from noon

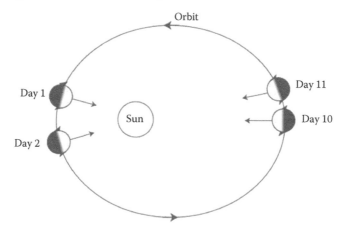

Figure 2.12 An imagined planet orbits the sun as it rotates, so it must rotate more than one complete revolution before the sun returns to its overhead position. The extra rotation is largest when the planet is nearest the sun.

to noon when it is nearest the sun and moving faster. For Earth, the effect is smaller than shown in Figure 2.12, and a "day" in January is only about 16 seconds longer than a "day" in July. This daily difference is hard to notice, but it builds up over a season. A sundial will be about 15 minutes slow in February and 15 minutes fast in October when compared to "clock time."

2.3.5.2 Melting Ice Caps   If the Greenland and Antarctic ice caps were to melt, Earth would get a little "fatter" as the melt water distributed evenly over the oceans. Days would be longer by a little more than half a second because of conserved angular momentum. The melting of floating ice would not contribute to the lengthening of a day.

2.3.5.3 Earthquake Correction   An earthquake that occurred in Japan on March 11, 2012, shortened the length of Earth's 24-hour day by 1.8 microseconds—at least theoretically. The quake allowed Earth to settle down a little, with higher places ending up lower. Then, conservation of angular momentum took over to speed the rotation and shorten the day. Other significant geologic events change day lengths by comparable amounts.

2.3.5.4 Earth-Moon Correction   The angular momentum arising from Earth's rotation is not quite conserved. The moon is slowing us down because it raises tides on Earth. As Earth rotates, the tides move ahead of the moon. As shown in Figure 2.13, when the moon pulls on these advancing tides, it exerts a torque that slows Earth's rotation ever so slightly. But, the difference adds up over time. About 900 million years ago, a day on a more rapidly spinning Earth was only 18 hours long.

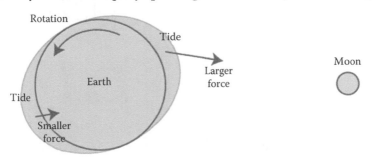

Figure 2.13   An exaggerated depiction of moon's torque on Earth as it pulls on the tides and slows Earth's rotation.

Conservation of angular momentum still applies to the Earth-moon system as a whole. The angular momentum lost by Earth is given to the moon. This causes the moon to drift away at less than a snail's pace of about 3.82 centimeters per year.

To compensate for the nonconstant rotation of Earth, a "leap second" is occasionally (midnight June 30 or December 30) added to the official clock time. This is a technological nuisance for people engaged in precise timing activities.

2.3.6 Astronomy

Conservation of angular momentum helps to explain many aspects of astronomy. The following are three of many examples.

2.3.6.1 Halley's Comet The peanut-shaped "icy snowball" about 15 kilometers long known as Halley's comet (also called Comet Halley) provides an extreme example of an elliptical orbit. When comets approach the sun, their emissions can make them clearly visible. The portion of the Bayeux tapestry of Figure 2.14 shows people observing

Figure 2.14 One small section (it is 70 meters long) of the Bayeux tapestry showing (in the upper right) Halley's comet in the year of the Battle of Hastings. In 1066, it was not called Halley's comet.

Halley's comet in 1066. Kepler's third law provides a measure of how far it moves from the sun. The cube of the sum of its closest and largest distances from the sun is proportional to the square of its period. Because this comet's period is about 75 years, one needs to find a number whose cube is the square of 75. A calculator gives approximately 17.8. Because the comet's closest approach is quite small, the farthest distance from the sun is about 35.6 times the distance from the sun to Earth. Its furthest distance is nearly 60 times its closest distance, so even though it may be invisible as it drifts to the outer regions of the solar system, we know that its top speed has slowed by a factor of 60 when it is at its greatest distance, off beyond ex-planet Pluto's neighborhood.

2.3.6.2 Solar System Formation   The currently favored theory for the formation of our solar system proposes a condensation process. Initially, a large region of low-density gas was pulled together by gravity. The original gas rotated slowly, but as it shrank, the conservation of angular momentum increased the rotation rate. Today, one notes that the sun and all the planets except Uranus rotate in about the same direction, with axes pointed roughly toward the North Pole star. Also, all the planets orbit the sun in the same direction. The initial angular momentum of the large gas cloud is responsible for all this angular momentum. There are some slightly troubling aspects of this model. The sun does not rotate very fast (once in about 25 days), but a day on Jupiter or Saturn is less than half an Earth day. Perhaps the sun has slowed because of energy loss. The transfer of angular momentum from the sun toward outer regions of the solar system can lower the total energy. The process may involve magnetic fields and turbulence.

2.3.6.3 Starquakes   Pulsars are special stars whose very rapid rotation rates can be detected by a periodic beam of radiation sent toward Earth. Sudden small increases in the rotation rate, called *pulsar glitches*, have been observed. "Starquakes," analogous to earthquakes, along with conservation of angular momentum may be the explanation for the glitches. Because no one has actually visited a pulsar, the physics may be much more complicated than a simple starquake.

2.3.7 Gyroscopes (and Tops)

The curiosity of the gyroscope of Figure 2.15 is easily stated. Why doesn't it fall? The answer is not so easy to find. Normally, an object tilts in the direction of the applied torque. But, when a gyroscope is rotating rapidly and the torque is small, the gyroscope axis moves in a direction perpendicular to the direction that would be expected from simple falling. This unusual motion is called *precession*.

The gyroscope makes sense if one follows the motion of individual points on the rotor and applies Newton's equation. Figure 2.16 simplifies the picture, but it can still look confusing. Various forces, velocities, and displacement are labeled with myriad arrows. The two arrows labeled "Force" represent the torque produced by gravity. The gray ring is initially spinning rapidly about the "Rotation axis." The direction of motion at the top and bottom of the disk are labeled "Velocity." Assume initially that the "Force" arrows are applied for just an instant to the top and bottom of the disk. Newton informs us that a force produces a change in velocity in the direction of the force. The "New velocity" at the top and bottom of the disk are denoted by the dotted lines. The changed velocity leads to changed positions, but the "Displacements" appear at the front and back of the disk because the rapid rotation transports the displacement away from the places where the forces were applied. The displacements mean the disk rotation axis

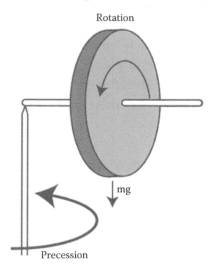

Rotation

mg

Precession

Figure 2.15 A gyroscope and the direction of precession produced by gravity.

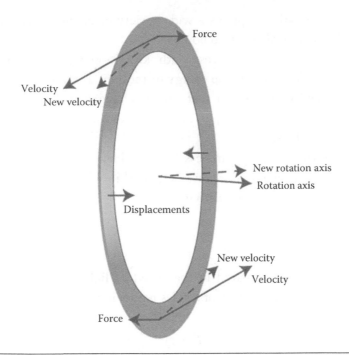

Figure 2.16 The forces, displacements, and velocities of the gyroscope shown in Figure 2.15.

is also changed to the dotted "New rotation axis." If the force is not just a pulse but continues to be applied, then the rotation axis continues to change, resulting in the precession.

Experiments show that the precession rate is proportional to the torque and inversely proportional to the angular momentum of the rotating disk. This makes sense. More torque produces faster precession. A larger disk radius with more mass is less responsive. If the rotation is faster, the time allowed for a point to move from the top to the front of the disk is decreased.

Gyroscope geometry is complicated, but it is important. Its generalizations are central to magnetic resonance and its many technological and medical applications.

2.4 Energy

Energy is conserved. That much is indisputable. The difficult part of energy conservation is keeping track. Energy can be transferred from one place to another, and it has more disguised forms than the "man of a thousand faces." But, treated with care, the fundamental law of energy conservation explains a wide variety of phenomena, ranging from athletic records to atom bombs.

Mechanical systems possess both kinetic energy and potential energy. Kinetic energy is the energy of motion. The potential energy is stored energy of position that can be changed to kinetic energy. Two potential energy examples are height above the ground and the compression of a spring.

Other forms of energy are less obvious. Heat is energy. Relativity relates energy to mass. These are described in Chapter 3 on thermal physics and Chapter 8 on relativity.

2.4.1 Kinetic Energy

For speeds small compared to the speed of light, a single formula describes the kinetic energy of a mass m moving with a speed v:

$$Kinetic\ Energy = \frac{1}{2}mv^2$$

Energy is measured in joules, named in honor of James Prescott Joule, a nineteenth-century scientist. A very fast runner can travel 100 meters in 10 seconds (or 10 meters per second). If the sprinter's mass is 80 kilograms, the formula tells us that the runner's kinetic energy is 4000 joules.

Kinetic energy should never be confused with momentum. Momentum has a direction; energy does not. The runner described here has momentum (mass times velocity). It is 800 kilogram-meter/second, but one must say which direction the runner is going to properly specify the momentum.

If a system is made of two or more parts, the rules for finding the total momentum and the total energy are completely different. When two objects are moving in opposite directions, the total momentum is obtained by subtracting. The total energy is the sum of the energies of each part, regardless of the direction of motion.

When Nicolas and Sonya push away from each other on the ice, their total momentum remains zero because they are moving in opposite directions. Their kinetic energies increase and Sonya's energy will be the greater if she is less massive than Nicolas.

2.4.2 Potential Energy of Height

Lifting an object increases its potential energy. When the object is dropped, the potential energy is changed to kinetic energy. The energy conservation examples that follow are related to the potential energy of height. This energy has a simple formula:

Potential Energy = mgh

The lifted object has mass m, $g = 9.8\,\text{meters/second}^2$ is the acceleration of gravity, and h is the distance the object is lifted.

2.4.3 Energy Conservation Examples

Example 2.4: Eagle and Turtle

An eagle wishes to eat a turtle, but it cannot penetrate its hard shell. So, the eagle shown in Figure 2.17 drops the turtle to the ground. The eagle understands that height is a form of potential energy, and dropping is the transformation of the potential energy of height into the kinetic energy of motion. The total energy is the sum of the kinetic and potential energies:

$$Energy = \frac{1}{2}mv^2 + mgh$$

Figure 2.17   Eagle dropping a turtle.

Before it is dropped, the potential energy of the 1-kilogram turtle dropped from 10 meters is *Energy* = *mgh* = 98 joules. Just before it hits the ground, all the potential energy has changed to kinetic energy, and *Energy* = *mv²* / 2 = 98 joules. Because half of 14 squared is 98, the eagle knows the turtle's speed will be 14 meters/second on impact. When the turtle hits the ground, its kinetic energy is changed into a variety of forms, including heat, sound, and dents in its shell. If a clever eagle determines that the speed must be doubled to break the shell, it will deduce that the drop height must be quadrupled. Eagles are unaware of the air resistance that will slow the turtle if it is dropped from a very large height.

Example 2.5: Car on Hill

A car runs out of gas, and the gas station is at the top of a hill. Can the car's motion carry it to the hilltop as shown in Figure 2.18? Energy conservation gives the answer. If the initial kinetic energy (1 / 2)*mv²* is greater than the potential energy needed to climb the hill *mgh*, then the car should make it to the hilltop. Reversing the turtle drop algebra, one finds that a speed of 14 meters/second allows the car to climb a hill 10 meters high. In the real world, air resistance and friction slow the car and change some of the kinetic energy to heat. The car might climb a hill only 9 meters high before it stops.

2.4.4 Work and Power

2.4.4.1 Work When a system is not isolated, its energy can change. For example, lifting a rock to a height *h* increases the rock's energy.

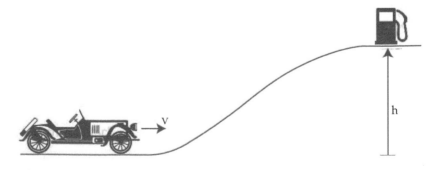

Figure 2.18 A car out of gas can make it up the hill to the gas station only if its kinetic energy of motion is greater than the potential energy needed to climb the hill.

This additional energy is the work done on the rock. Energy conservation means the person or machine lifting the rock loses an equal amount of energy. The exchanged energy is called the *work*. The change in potential energy *mgh* can be written as the product *mg* and *h*. Because *mg* is the force needed to lift the rock and *h* is the distance moved, the work done is the product of the force times the distance. Because work is a change in energy, work is also measured in joules.

A strong woman lifts 100 kilograms above her head to a height of 2.1 meters. The work done is

$$100 \text{ kilograms} \times 9.8 \frac{\text{meter}}{\text{second}^2} \times 2.1 \text{ meters} = 2058 \text{ joules}.$$

When she lowers the mass to the ground, the work done is negative because the mass returns to its original position and its original energy.

Lifting and lowering a weight results in no final change in the energy of the weight. But, chemical energy is changed to heat in the weightlifter. After lifting and lowering the weight, the man in Figure 2.19 will be warmer. The chemical energy comes from food.

Figure 2.19 The work done in lifting the weight is the mass times *g* times the distance up.

One Calorie (the dietary energy unit) is the same as 4184 joules. If energy conversion were perfectly efficient, 1 Calorie would supply enough energy to lift 200 kilograms more than 2.1 meters. In practice, the conversion of food energy to work is very inefficient. A person consuming 2000 Calories in a day cannot lift the 100-kilogram mass above his or her head 4000 times.

2.4.4.2 Power and Acceleration   Power is work (or energy) divided by the time it takes to perform that work. Its units (joules per second) are called watts. If a weightlifter takes 2 seconds to lift 100 kilograms 2.1 meters above his or her head, the weightlifter's power is 2046 joules divided by 2 seconds or 1023 watts. This is enough power to light 10 bulbs of 100 watts each. However, the bulbs would remain illuminated only for the 2 seconds of lifting. A horse can generate 1 horsepower or 745.7 watts. A strong person can replace a horse for only a very short time.

Your electric bill is a charge for energy. One kilowatt-hour is the same as 1000 *joules/second* × 60 *seconds/minute* × 60 *minutes/hour* × 1 *hour* = 3,600,000 *joules*. In principle, this is enough energy to lift three grand pianos (about a ton) to the top of the Empire State Building (381 meters). Electric utilities usually charge less than a quarter for each kilowatt-hour. If 25 cents worth of electricity can lift pianos so high, one wonders about electric bills. It appears that each month we are paying enough to lift many elephants to the top of the Empire State. Heat is the culprit that uses up so much electricity. This is described in Chapter 3.

Dividing work by time to obtain power gives an alternative expression for power. It is force multiplied by speed because speed is distance divided by time. This tells us the power needed to accelerate a car.

2.4.4.3 Fast Car   Assume that a Chevrolet Corvette can accelerate to 26.8 meters/second in 3.5 seconds. If the acceleration is uniform, the force applied to the 1500-kilogram car is a constant 11,490 newtons. The power needed to produce this acceleration is more complicated. It is the product of the force times the speed, so the largest power is needed at the end of the trip. At top speed, the required power would be 307,932 watts. This translates to an unrealistic 413 horsepower. So, what happened? Cars actually accelerate more rapidly at the beginning of the trip, and acceleration decreases as the speed increases. Measurements confirm greater acceleration at lower speeds. It is a better approximation to say that the power (rather than the force) delivered to the car is constant. That means the force delivered to the wheels decreases with the speed as the car's kinetic energy increases.

Fast animals can accelerate at slow speeds, but they cannot supply the power needed for high-speed acceleration. This explains why the greyhound can accelerate quicker than a Greyhound bus only at slow speeds.

2.4.5 Athletic Achievements

2.4.5.1 Pole Vault What is the highest achievable pole vault? A skilled vaulter gains maximum height by converting essentially all his kinetic energy into potential energy, as is sketched in Figure 2.20. The pole is the device used to facilitate the transformation of kinetic energy into gravitational potential energy. During the transformation, some of the energy is stored in the stressed pole. A bent pole is another form of potential energy.

No one can run much faster than 10 meters/second and achieving this speed is particularly difficult when carrying a long pole. A maximum speed implies a maximum kinetic energy that can be transformed, first into the bending of the pole and ultimately into the vaulter's height.

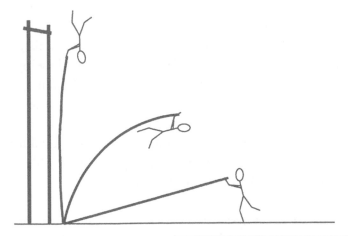

Figure 2.20 Time lapse pole vault sequence shows the conversion of kinetic energy to potential energy.

At first, it looks like physics has failed to describe the pole vault, so it is worth working out the numbers to find the errors. Assume the vaulter's mass is 80 kilograms and his maximum speed is 10 meters/second. This gives a kinetic energy of 4000 joules. At the top of the jump, all 4000 joules can be converted to the potential energy. Solving using $Energy = mgh$ gives the maximum achievable height: $h = 5.1$ meters. Again, this height does not depend on mass, so small and large people running at the same speed should be capable of achieving the same height.

Experimental reality in the form of well-documented pole-vaulting records presents a problem. In 1994, Ukrainian pole vault champion Sergey Bubka set a world record of 6.14 meters. That is 1 meter higher than the prediction, and it appears to violate physics. Where did the extra 1 meter come from? The apparent error illustrates two common problems of physics: misinterpretation of formulas and neglect of extra effects.

For the first error, the energy formula allows one to determine h, but this is the distance Sergey could jump above his initial position. The center of a standing person's body is about 1 meter above the ground, but the measured record is the distance between the ground and the bar. Good jumpers can arch their bodies so their body center is barely above the height of the bar, not 1 meter above the bar.

Regarding the second error, at the top of the pole vault, a person can push down on the pole, giving extra height. The work increases the energy, so energy conservation applies only if one includes the extra biomechanical energy stored in Sergey's muscles. These two errors mean a 6.14-meter vault is not a physical impossibility.

The pole vault is an example of material science allowing significant improvement in performance. In 1908, the Olympic pole vault record of only 3.71 meters (actually a tie) was set by Alfred Gilbert. Gilbert was also the inventor of the Erector Set. A wooden pole is not nearly as effective as today's bendy poles used to establish modern records. However, conservation of energy means sizable improvements on Bubka's 6.14-meter record are not possible.

2.4.5.2 Long Jump Physics (ballistics) also determines an upper limit for the long jump. The maximum range of a long jumper (or an arrow or a shot put) is achieved by aiming at a 45-degree angle

Figure 2.21 Long ranges are limited in part because a 45-degree angle takeoff is nearly impossible to achieve.

above horizontal. Then, the distance traveled, ignoring air resistance, is proportional to the square of the speed because higher upward speed keeps the jumper in the air longer and higher horizontal speed increases the range. The range is also inversely proportional to the acceleration of gravity. The formula is

$$Maximum\ range = \frac{v^2}{g}$$

Assuming again a maximum speed of 10 meters/second and using the acceleration of gravity $g = 9.8$ meter/(second)^2 means no jump should exceed 10.2 meters. This upper limit will be difficult to approach. As is suggested in Figure 2.21, it is hard to change enough of the horizontal motion into the vertical motion needed to maximize the range with a 45-degree angle liftoff. Also, wind resistance reduces the range.

The long jump is an example of technology playing a much less significant role than the pole vault. One can build a better pole, but not a better leg. As a result, record changes are less dramatic and less frequent. Jesse Owens jumped 8.13 meters in the 1936 Olympics. This record was unbroken for 25 years. Today, the world record long jump is still less than 9 meters.

2.4.6 The Pendulum, Harmonic Oscillators, and Resonance

2.4.6.1 Pendulum History A famous pendulum story about Galileo may be mostly true because it is attributed to Galileo himself. He was sitting bored in the Pisa Cathedral in 1602 when he noticed a brass chandelier swinging overhead. The remarkable aspect of the motion was the constant time needed for each back-and-forth swing (called the period). If the swing distance (called the amplitude) was doubled, the period stayed almost the same. Galileo used his pulse to check out the timing.

Galileo was surely not the first to have noticed that a pendulum's period is roughly constant, but he was the first to seriously propose using a pendulum for timing. By a strange coincidence, the earliest accurate experimental test of Galileo's ideas about falling objects used a pendulum to time the falls. This is one of many examples in the history of science for which a technical innovation (the pendulum) solidified a theoretical idea (constant acceleration). Galileo's observation of a pendulum eventually became the basic principle of clocks. Although Galileo and his son tried to make a pendulum clock, a practical example was not developed until 50 years later. Until the 1930s, clocks based on the pendulum were the world's most precise timekeepers, and various versions of these venerable clocks still decorate many homes.

2.4.6.2 Pendulum Physics The pendulum is another example of the interchange of kinetic and gravitational energies. Just as the speed of a pole vaulter determines the maximum height, the speed of a pendulum at the bottom of its swing determines how high it will go before it stops and falls back. As with the pole vault, the motion does not depend on the mass because both the kinetic and the potential energy are proportional to m.

The kinetic-potential energy interchange and the shape of the pendulum path explain its constant period. Assume a pendulum is given an extra "kick" that doubles its speed at the bottom of its arc. If the pendulum is moving twice as fast, the $mv^2 / 2$ kinetic energy formula is multiplied by four. When this kinetic energy is changed to potential energy, the pendulum will fly four times higher. Then, there is a geometry trick that saves the day. For the arc of a pendulum, four times the height is achieved by only doubling the amplitude, as is shown in Figure 2.22. If something goes twice as far at twice the speed, the time to make the trip (the pendulum period) is unchanged.

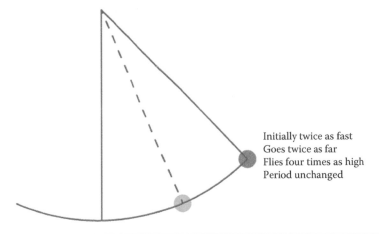

Initially twice as fast
Goes twice as far
Flies four times as high
Period unchanged

Figure 2.22 Doubled speed at the bottom doubles the amplitude of the swing. This means the pendulum period is almost independent of how high it swings.

2.4.6.3 Sine Wave A graph of the time dependence of the pendulum's position is a sine wave. Two sine waves with the same period but different amplitudes are shown in Figure 2.23.

2.4.6.4 Pendulum Generalization Examples of pendulum-like systems abound. They are called *harmonic oscillators*. Any system in which the kinetic energy is proportional to the square of the velocity and the potential energy is proportional to the square of a distance is a harmonic oscillator. A familiar harmonic oscillator is a mass suspended from a spring, as shown in Figure 2.24. The more a spring is stretched, the harder it is to stretch. The extra work needed to stretch a spring a long distance means the potential

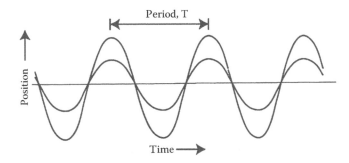

Figure 2.23 The time dependence of the position of a pendulum for small- and larger-amplitude swings. The period does not depend on the amplitude.

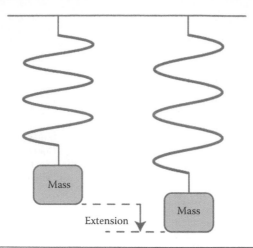

Figure 2.24 The motion of a mass suspended from a spring resembles the motion of a pendulum. Both are harmonic oscillators with a period independent of the amplitude.

energy is proportional to the square of the extension (rather than just the extension).

All graphs of harmonic oscillator motion (back and forth or up and down) have the sine wave shape of Figure 2.25, but they are distinguished by differing periods. The "natural frequency" is the inverse of the period. This frequency is measured in oscillations per second or *hertz*. Each string on a piano has its own natural frequency. The coffee sloshing back and forth in a cup has a natural frequency. Determining the natural frequency is often the first step in describing a physical system.

2.4.6.5 Equal Sharing of Kinetic and Potential Energies　There is symmetry in harmonic oscillators. The time averages of the kinetic and potential energies are identical and equal to half the total energy. One way to see the symmetry is to examine the sine wave curves that

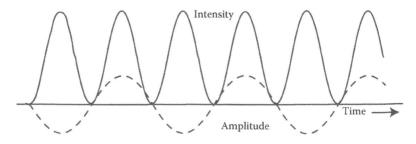

Figure 2.25 A sine wave (amplitude) and its square (intensity). The average of the squared sine wave is half its maximum.

characterize the motion. A sine wave and its square (the intensity) are compared in Figure 2.25.

The squared wave varies smoothly from zero to its maximum value and then returns to zero. The curve shape means the average value of the squared sine wave is half its maximum value. For a harmonic oscillator, the square of the displacement is proportional to the potential energy. At its maximum, the potential energy includes all the energy, so the average of the potential energy is half the total energy. The average of the kinetic energy is the other half of this total energy. As the pendulum swings, the energy is transferred back and forth between its kinetic and potential energy.

2.4.6.6 Resonance When a periodic force is applied to a harmonic oscillator, it responds dramatically when the frequency of the external force matches the natural frequency. This matching response is called *resonance*. Resonance allows one to determine the natural frequency. Many applications of resonance are seen in everyday life. (In practice, the terms *resonant frequency* and *natural frequency* are frequently interchanged.)

1. A radio receiver is tuned to a particular broadcast station by making the frequency of a harmonic oscillator built into the radio match the frequency of the radio signal.
2. Singing in a shower produces a particularly loud note when the resonant frequency of the air in the shower matches the note being sung.
3. A child on a swing "pumps" at a rate equal to the natural frequency of the swing-pendulum. This produces the large-amplitude swinging.
4. Water sloshing back and forth in a teacup, a bathtub, or the ocean can be approximately described by a harmonic oscillator. If the water is disturbed at a frequency that matches the natural frequency, the oscillations will be large. Your coffee can be spilled when the periodic bumps in the road match the coffee's natural frequency. An impressive example is the enormous, nearly 15-meter tides in the Bay of Fundy in eastern Canada, where the natural frequency of the bay matches the (nearly) 12-hour period of tides.

5. People on brisk walks often swing their arms at a rate that matches their pace. Resonance and easier swinging are achieved by bending the arms, with faster pace needing a sharper bend.

The key to resonance is shown in Figure 2.26. A string is attached to a pendulum, and each time the pendulum is at the bottom of its arc moving to the right, someone pulls on the string. The work done on the pendulum is the product of force and distance. Because the tugs on the string are timed to take place when the pendulum is moving in the direction of the force, each pull increases the energy. If the frequency of the periodic force did not match the resonant frequency, sometimes the pendulum would be moving against the force, and energy would be taken away from the pendulum.

Resonance applies to all harmonic oscillators, including masses suspended from springs and many electrical circuits. Even when a system is only approximately a harmonic oscillator, resonance can still occur, increasing the oscillation with an appropriately timed periodic force.

2.4.6.7 Does Resonance Collapse Bridges? In 1831, seventy-four men from the English Army Rifle Corps marched across the Broughton Suspension Bridge (Manchester, England). The resulting rhythmic swaying reportedly amused the troops, and they sang with the cadence and marched more vigorously. Then, the bridge collapsed and

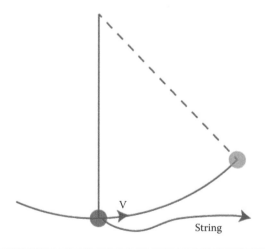

Figure 2.26 A pendulum attached to a string that can pull the pendulum back and forth.

injuries resulted. Because the collapse was blamed on resonance, it became the convention for British troops to "break step" when crossing a bridge. This rule soon became adopted for military marchers in many countries.

A few years later, the Angers Bridge in France collapsed under the weight of a battalion of soldiers, killing 200. In this incident, the wind was blowing the bridge about. It is theorized that the soldier's attempts to stay balanced may have enhanced the motion of the bridge at its resonant frequency.

In principle, the periodic force of marching could build up oscillations to the point of bridge collapse. However, bridge collapse stories should be approached with skepticism. It seems unlikely that the natural frequency of a bridge would match the stepping rate of soldiers.

2.4.6.8 Generalizations In many cases, resonant oscillations develop even though there is no apparent periodicity in the applied force. This generalized resonance requires a "feedback" that allows the force to acquire a resonant periodicity. Examples abound.

1. Musical instruments. The flute and violin are two examples for which there seems to be no periodicity in the applied blowing or bowing.
2. Singing. We can make sounds without shaking our bodies at the appropriate frequencies.
3. Microwave ovens produce a specific frequency even when the electric current source starting the oscillations is not periodic. The microwave is only one of many examples of electronic oscillators that use a nonperiodic energy source to produce a precisely defined resonant frequency.
4. Mechanical clocks oscillate at their natural frequencies even though the weights or springs that power the clocks are not exerting periodic forces.

A pendulum clock illustrates how the constant force of the mass leads to the periodic motion. The key is the interplay of the force and the oscillator that allows the force to be applied at the correct moment. In Figure 2.27, the clock escapement is arranged so a small returning force is applied to the pendulum after it has swung to the right. However, once the pendulum passes vertical alignment, the

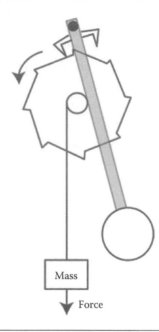

Mass

Force

Figure 2.27 A simplified pendulum clock escapement that gives the pendulum a forward "kick" once in each period.

escapement tooth to the left is disengaged. The tooth on the right stops the mass from falling. When the pendulum swings back past vertical, the process repeats and the pendulum gets another push.

In other cases, this "positive feedback" increases the amplitude at the natural frequency even though the feedback is difficult to identify. A whistle and the collapse of the Tacoma Narrows Bridge (both driven by moving air) are examples.

In some cases, the feedback can be related to peculiar properties of friction. The moving violin bow of Figure 2.28 catches the violin string and pulls it to the side. As soon as the string starts to slip past the bow, the friction decreases because the "kinetic friction" of objects moving past each other is generally smaller than their "static friction" when they are tied together. This means that the force is larger when energy is added to the string (force and velocity parallel) and smaller when energy is taken away (force and velocity antiparallel). The bow is pumping the violin string the same way a child pumps a swing, except the frequency is much higher.

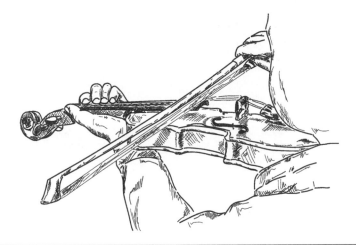

Figure 2.28   The bow sets the violin string into oscillations because static friction is larger than kinetic friction.

2.4.7 Energy and Momentum: Collisions and Billiards

Both the energy and momentum of an isolated system are conserved. The combination of the two conservation laws gives additional insight. A good example is the head-on collision of billiard balls. For simplicity, first assume the balls approach each other with opposite velocities (same speed, opposite direction), so the total momentum of the two balls is zero, and their kinetic energies are positive and equal.

2.4.7.1 First: Apply Conservation of Momentum   After the collision, the balls must again have opposite velocities. This is the only way to maintain zero total momentum. Because they cannot pass through each other, the ball initially moving to the right (left) rebounds to the left (right).

2.4.7.2 Second: Apply Conservation of Energy   The speeds before and after the collision are the same because the conserved total kinetic energy is proportional to the sum of the squared speeds.

Application of the two conservation laws means the balls interchange velocities when they collide head on. The velocity interchange is a general result that occurs even when the initial velocities are not equal and opposite, provided the balls have identical masses. In a billiards or pool game where a static colored ball is struck head on by a

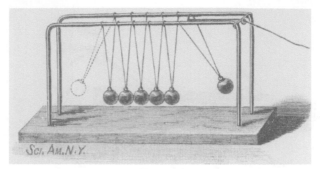

Collision Balls.

Figure 2.29 A Newton's cradle works because of energy and momentum conservation for each two-ball collision.

moving white cue ball, the cue ball stops and the colored ball acquires the velocity of the cue ball. The rolling energy of the balls was ignored, but this does not change the velocity interchange. In practice, the velocity interchange is not perfect because friction cannot be ignored.

Collisions need not be restricted to simple two-ball problems. In principle, one could place several pool balls in a line, strike the first one and see a sequence of velocity interchanges until the last ball in the line acquired the speed of the first ball. It is difficult to align pool balls with sufficient accuracy to produce this chain reaction. However, the alignment is guaranteed in the toy of Figure 2.29 called Newton's cradle; the momentum of the first ball is transferred to the last ball through a rapid sequence of head-on collisions. In the end, all the intervening balls appear to be unaffected.

Scattering is more complicated when collisions are not head on, and approximations that ignore friction are not very accurate. This is one of the many reasons why experts in physics usually do not excel at billiards.

2.4.8 Escape from Gravity

A bullet shot straight up at 11.2 kilometers/second would escape Earth and never fall back, provided air resistance is ignored. One need not worry about lost bullets because 11.2 kilometers/second is about 30 times the speed of sound and much faster than any gun can produce. This "escape velocity" is determined by equating the kinetic energy of the bullet to the potential energy associated with the gravitational attraction of Earth. Gravity weakens with increasing altitude, so only finite energy is needed for an escape.

An object with only half the kinetic energy for escape is still moving very fast. Its energy is sufficient to orbit Earth. The two-to-one relation between "escape energy" and "orbit energy" is quite general. If Earth were to be given a giant kick from behind that doubled its kinetic energy, we would escape from the sun. There is no conceivable force that could do this without demolishing everything.

The same relation between orbit energy and escape energy applies for an electron orbiting the proton of a hydrogen atom. Doubling the kinetic energy of the electron frees it from the nucleus and ionizes the hydrogen atom. The microscopic energies of atomic physics mean atomic ionization is important and commonly observed in chemistry.

The analogy between the Earth–sun system (ruled by gravity) and the hydrogen atom (ruled by electrostatics) makes sense because the two forces share the inverse square law. Quantum mechanics complicates the atomic physics, but the solar system analogy remains helpful.

3

THERMAL PHYSICS

3.1 Introduction

The beauty of thermal physics lies in its universally applicable principles. Examples described here hint at the broad reach of thermodynamic and statistical principles.

A particularly important application of thermal physics relates to energy. When we say the world "consumes" energy, we mean the energy is transformed into a form that makes us happier, by powering a car, lighting the dark, or charging a cell phone.

3.1.1 Energy Sources

The sun is the ultimate source of nearly all our energy (Figure 3.1). Conversion of the sun's energy into useful forms is a uniquely human accomplishment. Fire is the earliest example. A tree uses solar energy to produce wood (mostly from water and carbon dioxide). When we burn wood (or, more generally, "biomass"), we are recapturing the sun's energy that is stored in the chemical energy of the tree.

The modern world consumes huge quantities of energy. Using wood alone would quickly deforest the planet. Today, most of our energy is obtained from the burning of coal, oil, and natural gas. The

| Sun | Plant | Coal | Heat | Electricity |

Figure 3.1 Most of our energy comes from the sun, but the path from sunlight to electricity is complicated.

sun is also the ultimate source of these fuels. Buried plant and animal life eventually become fossil fuels.

Some solar energy can be captured without burning. Photovoltaic devices (Chapter 7) convert sunlight directly into electricity. Arrays of mirrors can concentrate the sunlight to produce high temperatures. Hydroelectric power is also ultimately derived from the sun, which evaporates the water and leads to rain and rivers. Wind is produced by the sun's uneven heating of the planet. Despite significant increases in these special energy sources, they represent a small contribution to our total energy budget.

Nuclear energy does not come from the sun. Radioactive elements like uranium produce energy as they decay. When properly managed, this decay results in usable heat. Radioactivity also heats the interior of the planet. In some places, notably in Iceland, this geothermal energy is sufficient to be of use. Nuclear power does not add carbon dioxide to our atmosphere. However, there are practical and political problems associated with the disposal of nuclear waste. Nearly 10% of energy in the United States is obtained from nuclear power.

3.1.2 Problems

Burning anything produces carbon dioxide as well as water and smaller amounts of other chemicals. Fossil fuel combustion has increased the carbon dioxide in our atmosphere and oceans. These well-documented changes are certainly a concern. It is sensible to believe that these alterations of our planet lead to climate changes and other environmental effects. Measurements tell us the global temperature is increasing. The character and extent of climate changes are hard to understand and to predict. This is one reason a causal relation between human activity and alterations in our climate have been questioned.

The world is addicted to energy. In addition to supplying the comforts of everyday life, energy's role in fertilizer production helps to feed a growing population. The consumption of fossil fuels continues to increase, and the resulting increase of atmospheric carbon dioxide is accelerating. Reserves of coal, oil, and natural gas are finite. Fuels are expensive, and their importation affects a nation's wealth.

The problems of climate change and the limited supply of expensive energy resources have led many to call for "energy efficiency." Improved gas mileage in cars, more wind generation, and various other sensible changes can make a difference. However, we should realistically face the enormous magnitude of the problem. If the United States could make a miraculous change and use only energy sources that do not add carbon dioxide to the atmosphere (wind, hydro, nuclear, photovoltaic, etc.), the worldwide effect would be modest. In less than a decade, the saving would be gobbled up by ever-increasing fossil fuel use in other countries. In my opinion, the steps being taken to avert an energy-climate catastrophe are insignificant. The catastrophe is inevitable.

3.1.3 Energy Conversions

Wind, hydropower, and photoelectric devices (presently less than 10% of total energy production) can be transformed directly to electricity. For example, hydropower can convert falling water to electricity with nearly 90% efficiency.

Most energy sources used today generate heat. A steam engine and an automobile engine are examples of the conversion of this heat to mechanical energy. The sad fact is the inherent inefficiency in this energy conversion. This is not an engineering problem; it is a consequence of basic thermal physics. So, if one really wants to understand energy, one must know the principles of thermal physics described in the material that follows.

3.2 Ideal Gas

A box filled with helium atoms (Figure 3.2) would appear to be a particularly simple gas. Essentially all helium atoms are identical, with two electrons hovering around a compact nucleus composed of two protons and two neutrons. These atoms are submicroscopic spheres with radii of about 1/10 nanometer. (It takes 1 billion nanometers to make a meter.) At room temperature and pressure, the average distance between the atoms is 50 to 100 times the atomic radius, and an atom typically travels more than 1000 atomic radii before it collides with another atom. Helium atoms have weak chemical interactions and no complicated internal structure. All this means helium gas at room temperature and pressure is a nearly "ideal" gas made of independent spherical atoms that fly about freely and only occasionally collide.

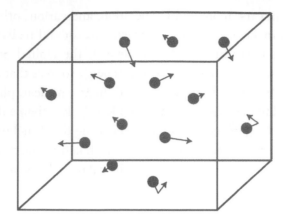

Figure 3.2 Helium atoms in a box. Their kinetic energy is a form of heat. Atoms bouncing from the walls produce pressure.

The helium gas atoms move rapidly. This motion means the gas has kinetic energy. At room temperature, the typical atomic speed is about 1250 meters/second. Because the atomic mass is only 6.65×10^{-27} kilograms, the energy of an individual atom is exceedingly small, but a cubic meter of this gas at room temperature and pressure contains 2.5×10^{25} atoms. Thus, the energy of a cubic meter of helium gas is more than 20,000 joules. This surprisingly large energy is about the same as the kinetic energy of five people running at top speed (10 meters/second).

Because atoms are invisible, it is an easy mistake to ignore the atoms' considerable energy. The total energy (mostly kinetic energy for helium gas) is called *heat*. The equivalence of heat and energy is the general principle called the *first law of thermodynamics*.

3.2.1 First Law of Thermodynamics

Heat is a form of energy. When heat is taken into account, total energy is conserved.

The first law of thermodynamics also applies to oxygen (and everything else). Oxygen gas is made of diatomic molecules with two atoms sticking together. For oxygen and other molecular gases, there is heat kinetic energy associated with molecular rotations that is absent in helium.

Heat is also potential energy. At high temperatures, a significant amount of the heat in oxygen molecules is stored in vibrations because the molecular bond is like a spring. Vibrations share equal amounts of kinetic and potential energy.

The heat energy in solids and liquids is shared among a variety of forms of kinetic and potential energy. One can gain insight into the physics of materials by measuring the amount of stored heat and its changes with temperature.

3.2.2 Temperature

Temperature is not the same as heat. However, the two are related because energy is always required to raise the temperature. An ideal gas like helium is particularly simple because the energy and temperature are proportional to each other. History dictates that temperature be measured in degrees, but heat is an energy measured in joules. Thus, one needs a constant k (called the Boltzmann constant) to change a temperature T to an energy. The following is the result for helium gas:

$$\frac{3}{2} k \times Temperature = average\, atom\, energy$$

The 3 is associated with three-dimensional space, and the 2 reflects the ½ in the kinetic energy formula: $(1/2)mv^2$. This relation between temperature and energy demands that temperature T be measured with respect to "absolute zero," so $T = 0$ is the temperature at which the energy can go no lower. The kelvin temperature scale is commonly used, where $T = 273\, K$ is the freezing point of water and $T = 373\, K$ is the boiling point of water. Other characteristic kelvin temperatures are shown in Figure 3.3.

The sun's temperature indicated in Figure 3.3 is only an approximation. Sun spots, turbulence, or magnetic storms mean the sun's surface temperature is not clearly defined.

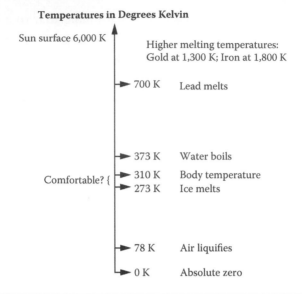

Figure 3.3 The kelvin temperature scale with some characteristic temperatures. Humans are comfortable only in a narrow temperature range.

3.2.3 Pressure

When a liquid or gas is compressed, it pushes back with a pressure and a force related by

$$Force = Pressure \times Area$$

The ubiquitous atmospheric pressure is produced by Earth's gravity compressing the air. This force is the weight of a column of air extending up to the top of the atmosphere.

No one can suck water through a straw that is more than 10.3 meters high or siphon water over a 10.3-meter hill. Water rises in a straw only because atmospheric pressure is pushing it up. Similarly, the atmospheric pressure can elevate mercury only about 0.76 meters.

For a given cross-sectional area, three columns have the same mass and weight: 10.3 meters of water, 0.76 meters of mercury, and the atmosphere all the way up to the top.

Pressure increases under water. At 10.3 meters below the surface, the total pressure is double the atmospheric pressure. Deep divers must use care when returning to atmospheric pressure because high-pressure gases can be absorbed into body fluids. Rapid decompression is a little like opening a carbonated drink. Bubbles forming in the body are painful and sometimes lethal.

It is a strange comment on the relationship between money and science that the one person who descended to the deepest ocean (11 kilometers) is James Cameron, famous vegan and movie person. An ultrastrong submarine was needed to shield him from more than 1000 times atmospheric pressure.

3.2.3.1 Gas Pressure The pressure of a gas such as helium is a result of the atomic motion. When a gas is placed in a container, the atoms bounce from the walls and push them out. This force is proportional to the product of the number of collisions per second and the momentum change produced by each collision. Faster atoms mean more collisions with the walls. Faster atoms hit the walls harder (a larger momentum change). The product of momentum mv and velocity v (with a factor of ½) is the kinetic energy $(1/2)mv^2$, and the kinetic energy is proportional to the temperature. Filling in the details yields

$$Pressure = k \times Temperature \times [Atomic\ density]$$

The k is the same Boltzmann constant that related energy and (kelvin) temperature. The atomic density (number of atoms in a cubic meter) appears in this pressure equation because more atoms produce more frequent collisions with the walls. Pressure is proportional to the temperature only for an ideal gas. When atoms are crowded together to make a solid or a liquid, the pressure is very different.

3.2.3.2 Pressure and Work For many applications, pressure is the method by which heat is changed to mechanical energy. Work (energy change) is force times distance. The thermodynamic generalization of work is pressure multiplied by the volume change. The steam engine of Figure 3.4 is an example of the use of pressure to do work. Car engines use the pressure of burned gas to push the cylinders. These and other practical examples of the transition of heat to mechanical work are technically complicated. Some aspects of the steam engine are described in Section 3.4 (on phase transitions).

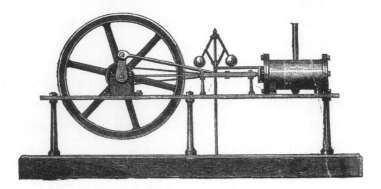

Figure 3.4 Steam engine. Pressure difference between inside and outside the cylinder pushes a piston back and forth and turns the wheel.

3.2.4 Heat Flow

Suppose one has a container with two compartments separating hot and cold helium. The gases are allowed to mix. One expects the comparatively fast atoms from the hot gas to lose energy (on the average) and give that lost energy to the slower atoms that comprise the cold gas. This makes sense because when two atoms collide, it is usually (but not always) the case that the faster atom slows and the slower atom speeds up. This unsurprising result generalizes to the *second law of thermodynamics*.

3.2.5 Second Law of Thermodynamics

3.2.5.1 Heat Flows from Hot to Cold The almost-obvious second law (see Figure 3.5) has broad applicability, but there are some provisions.

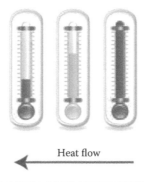

Heat flow

Figure 3.5 The sensible flow of heat from hot to cold is mandated by the second law of thermodynamics.

For example, one cannot use the second law to prove that refrigeration is impossible. Heat is guaranteed to flow from hot to cold only if external forces and energies are absent. For an isolated system, like a mixture of cool and hot gases, the second law applies.

A proof of the *second law of thermodynamics* is more difficult than one would expect. One must deal with averages and statistics, and the second law appears to apply only for a large number of atoms. Uneasy feelings about the second law have been around for a long time. Around 1870, James Clerk Maxwell contributed to this unease when he proposed what appears to be a counterexample:

> Now let us suppose that ... a vessel is divided into two portions, A and B, by a division in which there is a small hole, and that a being, who can see the individual molecules, opens and closes this hole, so as to allow only the swifter molecules to pass from A to B, and only the slower molecules to pass from B to A. He will thus, without expenditure of work, raise the temperature of B and lower that of A, in contradiction to the second law of thermodynamics.

The "being" of Figure 3.6 became known as the "Maxwell demon." This demonic doorkeeper must be remarkably perceptive with quick responses. Many clever and famous people, including Richard

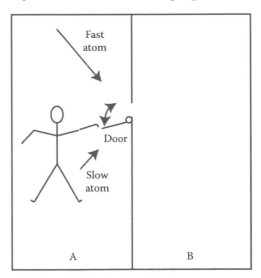

Figure 3.6 Maxwell demon adjusting the door so the fast atom can pass but the slow atom is blocked. Repeating this stunt facilitates the flow of heat (energy) from cold to hot, violating the second law of thermodynamics.

Feynman, have concocted a variety of machines that attempt to repli-cate the accomplishments of Maxwell's demon. Most have concluded that Maxwell's demon and his many cousins are doomed to fail in their attempts to violate the *second law of thermodynamics*, but contro-versies about deeper meanings remain.

3.3 Entropy

Unlike pressure, temperature, or energy, entropy is not intuitive. It has two definitions. One is based on information, and the other is thermodynamic. The interplay of these definitions makes entropy as important as it is mysterious.

3.3.1 Entropy's Dual Definition: Information and Thermodynamics

3.3.1.1 Information Definition The less one knows about a system, the greater its entropy. Helium gas is a good example. We do not know the positions or velocities of the atoms, so the gas has entropy. One would need to gather a great deal of information to gain com-plete knowledge of the many atoms.

The amount of information needed to know everything about a system provides the information-based definition of entropy. This information could be obtained from a series of answers to "yes-no" questions. As a simple example, imagine a gas composed of only two atoms, one helium atom and one argon atom. The first ques-tion could be: Is the helium atom in the top half of the box? Many more questions would be needed to nail down the position of the helium atom to within atom-scale dimensions. Another set of ques-tions could locate the argon atom. Even more questions could deter-mine the velocities of the two atoms. The entropy is defined by the minimum number of cleverly worded questions needed to determine all the unknown quantities.

$$Entropy = 0.69 \cdot k \cdot (Number\ of\ questions)$$

A version of this fundamental definition of entropy is printed on Ludwig Boltzmann's tombstone. The k in this equation is the same Boltzmann constant used to relate energy and pressure to temperature.

The 0.69 is the natural logarithm of 2. It appears in the formula because the questions posed have two answers (yes and no).

The equation says a perfectly ordered system has no entropy because no questions would be needed to determine its properties. A disordered gas has a great deal of entropy. Entropy will be larger if there are more atoms or a larger container. Higher temperature produces more entropy because the higher average speed requires more questions.

3.3.1.2 Thermodynamic Definition The thermodynamic definition of entropy relates the change in entropy to the heat added to a system.

$$Entropy\ increase = \frac{Heat\ added}{Temperature}$$

The two definitions of entropy make sense only if one can prove that they are equivalent. As a shortcut around a proof of equivalence, one can roughly justify the thermodynamic definition of entropy by doubling the volume of a box containing N atoms (at a fixed temperature). The doubled volume means one extra yes-no question is needed to determine each atomic position. According to entropy's information definition, that means

$$Entropy\ increase = 0.69 \times N \times k$$

Pressure exerts a force on the walls of the container, so if the container expands the gas does work. The work obtained from the expansion must be supplied by an outside heat source or the gas would cool. The heat needed to keep the temperature constant is

$$Heat\ added = 0.69 \times N \times k \times Temperature$$

For this case, the natural logarithm of 2 appears because the ratio of final to initial volume is 2:1.

It is not a coincidence that the entropy change is the ratio of the heat added divided by the temperature. It must be this way if the two definitions of entropy are equivalent.

Entropy can explain the failure of Maxwell's demon. After the demon decides whether to let a molecule pass from A to B, it must forget that decision before it deals with the next molecule knocking on the door. The information definition of entropy means forgetting increases entropy. The thermodynamic definition means an increase in entropy is accompanied by added heat. Thus, the demon (like an air conditioner) must be supplied energy to perform its selection process, and Maxwell's demon is no more than a fancy air conditioner.

3.3.2 Fundamental Property of Entropy

The entropy of an isolated system never decreases.

Entropy is associated with disorder and a lack of information. The fundamental property means that isolated systems never spontaneously simplify. This almost-obvious result is a foundation of thermal physics.

The fundamental property means a system with less than the maximum entropy may evolve in time to a state with larger entropy. When entropy achieves its maximum value and no further increase is possible, one says the system is in "thermal equilibrium."

Helium is an example that shows how entropy maximization determines the properties of a system. Assume all the helium atoms are initially constrained to occupy only the left-hand side of a container. This is not a state of maximum entropy. When released, the helium atoms become uniformly distributed, with distribution to the right as likely as left for any atom. The entropy is increased because one additional question, "Is the helium atom in the left- or right-hand side of the box?" must be asked of each atom. This means the entropy increase is $0.69 \, Nk$, where N is the number of helium atoms and k is again the Boltzmann constant.

Black holes appear to challenge the rule that entropy can only increase. The entropy of objects absorbed by black holes would appear to disappear. One can assign entropy to black holes even though this entropy cannot be observed. Suffice it to say, many aspects of black holes remain mysterious.

3.3.2.1 Heat Flowing from Hot to Cold An example of an entropy increase is shown in Figure 3.7: Heat flows from a hot region of a system to a cold region. Heat lost from the hot region decreases entropy, but the heat arriving in the cold region increases the entropy. The final result is an increase in the total entropy. This follows from the thermodynamic definition of entropy. It also follows from Figure 3.7

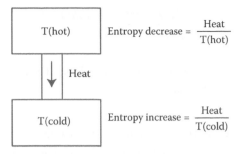

Figure 3.7 Heat flowing from hot to cold increases the total entropy because the low-temperature increased entropy is larger than the high-temperature entropy decrease.

that heat flow from cold to hot is impossible because this would decrease the total entropy. The unavoidable increase in entropy is an alternative expression of the second law of thermodynamics (heat flows from hot to cold).

3.3.2.2 Changing Heat to Work   In thermal physics, "heat" is associated with entropy, but "work" is mechanical or electrical energy that is free of entropy. It is not possible to completely change heat into work because this would decrease entropy. Entropy can only be preserved if some heat is wasted. The best possible heat engine wastes the minimum amount of heat by creating no additional entropy. An example ideal heat engine is shown in Figure 3.8.

The minimum possible waste heat is achieved when the entropy decrease at $T(hot)$ is exactly canceled by the entropy increase at $T(cold)$. Equating these entropy changes determines the minimum waste heat ratio in terms of the two temperatures.

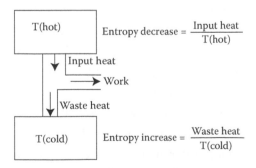

Figure 3.8 Heat flowing from hot to cold can generate work. The maximum work corresponds to no change in the total entropy.

$$\frac{Waste\ heat}{Input\ heat} = \frac{T(cold)}{T(hot)}$$

It is no surprise that no work can be obtained if there is no difference between $T(hot)$ and $T(cold)$.

All power plants using heat are subject to this efficiency limitation. It does not matter if the heat comes from coal, oil, natural gas, or nuclear fission, a significant fraction of the heat must be discarded. This waste heat is often given the unsympathetic name *thermal pol-lution*. There is good reason for locating big power plants near large bodies of water because the water can be used as the low-temperature reservoir. Modern power plants generate huge amounts of energy, so correspondingly large amounts of waste heat must be dumped into the low-temperature reservoir. In some cases, this significantly increases the water temperature. There are reported cases of alligators moving to colder climates near power plants because of the heated water.

The tragic March 2011 earthquake destruction of the nuclear power plant in Fukushima, Japan, polluted ocean waters with radio-activity. With hindsight, locating a nuclear power plant near an ocean may seem foolish, but the need for efficient cooling is a fundamental requirement of all power plants. All 50 of Japan's nuclear reactors lie near the coast.

An optimistic example shows how much energy is wasted. A power plant burns fuel to produce a $T(hot)$ of 600 kelvin. It uses a lake at 300 kelvin for $T(low)$. The 2:1 ratio of the temperatures means half the heat would be wasted by ideally efficient energy conversion. Typical power plants burning fossil fuels are less efficient, converting no more than 40% of the heat into electricity. Higher efficiency can be achieved with larger values of $T(hot)$, but this is both expensive and dangerous. One can do slightly better with "cogeneration," by which the waste heat is used to heat homes.

One of many ideas for obtaining energy at no environmental cost proposes using the ocean. Water from the deep is the low-temperature reservoir and surface water in the tropics is the high-temperature reservoir. Unfortunately, the ideal efficiency limit makes this plan unappealing. Using an optimistic $T(cold) = 273\ K$ (the freezing temperature) and $T(hot) = 298\ K$, the ratio of waste heat to input heat is nearly unity.

$$\frac{Waste\ heat}{Input\ heat} = \frac{273}{298} = 0.916$$

A proposed project in which at least 91.6% of the heat is wasted is not likely to achieve widespread support.

In some places, subterranean water is hot enough to allow the generation of electricity without burning fuel. In many cases, especially in Iceland, moderately hot geothermal water is used directly to heat houses. This avoids the efficiency limitation.

It does not make good physics sense to use electricity to heat homes because the conversion of heat to electricity is intrinsically inefficient. In principle, it should always be more economical to use combustion heat directly.

3.3.2.3 Cooling   You cannot cool your kitchen by opening the refrigerator door. The cooling when the door is opened is accompanied by heating of the coils in the back of the refrigerator. The final result is a hotter kitchen because all the electrical energy used to run the refrigerator becomes heat. An air conditioner is different from a refrigerator because the heated part is placed outside the home.

One can describe idealized cooling as running a heat engine in reverse. The most efficient cooling system creates no entropy because extra entropy means extra heat. The cooling engine illustrated in Figure 3.9 is essentially the same as the heat engine in Figure 3.8 except the directions of the energy flows are reversed and the terminology is changed to represent cooling.

For air conditioning, the $T(hot)$ region is outside the house, and $T(cold)$ is the inside temperature. The engine in Figure 3.9 describes the pumping of heat from inside to outside, with some additional heat flowing out because of the work needed to pump the heat. The waste heat is no longer the undesirable quality. It is the heat removed by the air conditioner. The work is no longer the desired output of a power plant. It is the energy cost of air conditioning. A relabeled version of the formula for the ratio of waste heat to input heat says

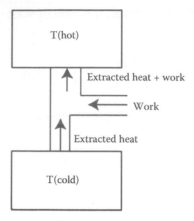

Figure 3.9 Running the heat engine in reverse allows refrigeration. The names are changed, but the physics is essentially the same.

$$\text{Heat removed from cold} = \frac{T(cold)}{T(hot)}(\text{Heat delivered to hot})$$

This provides a familiar result. The hotter it is outside, the less heat can be removed by an air conditioner. As an example, assume the inside temperature is a comfortable 295 K (kelvin temperature), and outside it is an uncomfortable 310 K. These numbers mean the heat removed from cold is 295/310 = 95% of the heat sent outside. The extra 5% going outside is the energy needed to run the cooler. Ideally, this means air conditioning would remove nearly 20 times as much heat as is used to power the air conditioner. This optimistic number is the limit imposed by thermodynamics. In practice, air-conditioning systems do not approach this 20:1 ratio.

One can ask about the extreme limit. Suppose one wanted to cool a system to absolute zero. There is a problem because "heat removed from cold" vanishes when

$$T(cold) = 0$$

This suggests the *third law of thermodynamics*.

3.3.3 Third Law of Thermodynamics

Nothing can be cooled to a temperature of absolute zero.

3.4 Phase Transitions

The boiling of water is but one example of a phase transition. It is of special importance because of its application to steam engines. Other examples of phase transitions are described at the end of this section.

3.4.1 Boiling Water

Water, ice, and steam (water vapor) are all made of water molecules consisting of two hydrogen atoms tightly bound to an oxygen atom. When water boils, it undergoes a "first-order" phase transition." Although the water molecule of Figure 3.10 remains intact, there are many differences between liquid and gaseous phases of water.

Steam is a gas. As with helium gas, the molecules are separated by relatively large distances, and molecules travel many times the molecular radius between collisions. In liquid water, the molecules are much closer together, essentially touching each other. The large atomic spacing of steam means that at atmospheric pressure and the boiling temperature (373 K), the gaseous form of H_2O takes up 1700 times as much space as the liquid form. Because the liquid water molecules lie in close proximity, it is difficult to compress water.

Water molecules attract each other, and separating the molecules is difficult. Because of this attraction, it takes 5.4 times as much energy to boil water as it does to heat water from the freezing temperature (273 K) to boiling temperature (373 K). This is familiar to anyone heating a pot on the stove. The time needed to start water boiling is much less than the time needed to boil all the water away.

The key to the liquid-gas phase transition is the interplay between energy and entropy. At the boiling temperature, there is sufficient energy to make the transition from lower-energy water to the larger-entropy steam.

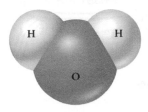

Figure 3.10 Water molecule.

3.4.2 Steam Engines

Steam engines use the liquid-gas phase transition to advantage. Heron's earliest steam-driven device of Figure 3.11 boiled water in a pivoted spherical container mounted on an axis. The steam was ejected through two spouts oriented to rotate the sphere. The steam outlets acted like jets, with the momentum of the ejected steam pushing the opposite sides of the sphere in opposite directions.

Heron's device would spin, but it never generated any usable energy, so it was not exactly a steam engine. The earliest commercially viable versions of steam engines worked differently (Thomas Newcomen 1712). Steam was placed in a cylinder with a movable piston. The steam pressure equaled atmospheric pressure when the plunger was extended. Then, the cylinder was cooled. Almost all the water vapor turned to liquid water, which took up almost no space. The result was a drastic drop in the cylinder pressure. Atmospheric pressure pushed the plunger into the cylinder with a force that could do mechanical work. Steam engine design has continued to evolve since the earliest version. The first steam engines were used primarily to pump water from mines. Later engines powered a wide range of industrial machinery with increasing efficiency. The steam engine was truly the workhorse of the Industrial Revolution.

Figure 3.11   Heron's steam sphere.

One might think steam engines are historical relics replaced by internal combustion engines and electric motors. This is not the case. Most (around 80%) of our electrical energy is derived from the burning of fuel to generate steam. The steam is ejected through nozzles directed at turbine blades. The lift force of the fluid rapidly spins the turbine. The energy of the rotating blades is changed to electrical energy, as described Chapter 4 (Electromagnetism).

3.4.3 Other Phase Transitions

The melting of ice is another first-order phase transition. The physics is different because the transition of solid to liquid generally involves only small volume changes. The molecules in the liquid stay close together when the ordered structure of the solid disappears and is replaced by the randomness of the liquid. Molecules can move about in a liquid, but they are locked in place in a solid. Thus, ice is hard and water is not.

Another example of a first-order phase transition is seen in the transition from solid directly to gas. This occurs for carbon dioxide: The solid "dry ice" changes directly to gas. Liquid carbon dioxide can only be produced under high pressure.

Second-order phase transitions are gentler than first-order transitions such as melting and boiling. A second-order example is iron's magnetism. The magnetism vanishes at the transition temperature, but it does so smoothly, with the magnetism becoming weaker and weaker as the transition is approached. This is quite different from the first-order transition of melting. Ice does not become softer and softer as it is warmed to the melting temperature. Other second-order phase transitions are associated with superfluidity and superconductivity.

3.5 Random Walks

Random walks like that shown in Figure 3.12 may seem like a walk into irrelevance. But, you never know what can be found when traveling off the beaten path. Diffusion, conduction, Brownian motion, and the equipartition theorem are all related to the random walk.

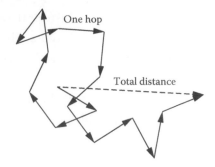

Figure 3.12 A random walk in two dimensions. Many hops typically lead to a total distance that is proportional to the square root of the number of hops.

3.5.1 A Random Walk in One Dimension

Phyllis, a frog, plans an adventure. Her jumping distance is $HOP = 1$ *meter,* so if she hops once every second for 15 hours a day, she should still be able to circle the world in 800 days. However, frogs have a lousy sense of direction and a poor memory. After each hop, Phyllis flips a coin. Heads means she hops east, and tails produces a hop to the west. This poor frog ends up hopping back and forth in an unpredictable random walk. It is a good bet that Phyllis Frog will not go around the world in 800 days.

Because coin flips are random, only the average properties of the frog trip can be described. If the coin is unbiased, the average distance Phyllis travels to the east is zero because travel to the west is just as likely. If the frog distance to the east is denoted x, then $average(x) = 0$.

The square of Phyllis's progress after N hops (averaged over all paths) is a more useful way to characterize her random walk. After one coin flip, $x = +1HOP$ or $x = -1HOP$. Because the square of both possibilities is positive, $average(x^2) = 1(HOP)^2$ for one flip. For two coin flips, the four possible paths, with plus and minus corresponding to heads and tails, respectively, are $(+,+),(+,-),(-,+),(-,-)$. The possible values of x/HOP for the two-flip trip are 2, 0, 0, -2. The average squared distance is obtained by squaring each of the four values of x, adding them, and dividing by four. The result is $average(x^2) = 2(HOP)^2$. For a three-hop trip, the bookkeeping becomes tedious, but not impossible. The eight possible coin-flip sequences are $(+, +, +)$, $(+, +, -)$, $(+, -, +)$, $(-, +, +)$, $(+, -, -)$, $(-, +, -)$, $(-, -, +)$, $(-, -, -)$.

Taking the average of the squares means one must compute

$$\frac{1}{8}(9+1+1+1+1+1+1+9)d^2 = 3(HOP)^2.$$

A pattern is emerging.

One hop: $\text{average}(x^2) = 1 \times (HOP)^2$

Two hops: $\text{average}(x^2) = 2 \times (HOP)^2$

Three hops: $\text{average}(x^2) = 3 \times (HOP)^2$

If the pattern works for N = 1, 2, 3, maybe it works in general. The proof of the generalization is interesting only for people who have already seen it, so take it as a given that

$$N \text{ hops:} \text{average}(x^2) = N \times (HOP)^2$$

A useful estimate of how far Phyllis Frog has traveled after N hops is the square root of $\text{average}(x^2)$. This gives the magnitude of the typical distance moved.

$$DISTANCE \approx \sqrt{N} \times (HOP)$$

The square-root dependence on the number of hops N is shown in Figure 3.13. The $\approx$ symbol means roughly the same, but only on the average. After N hops, she might end up where she started. She might also go much further.

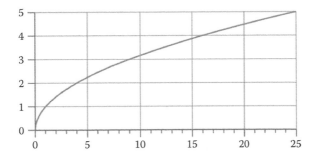

Figure 3.13 Square root. The horizontal axis is the number of hops. The vertical axis is the typical distance (with $HOP = 1$). This distance grows more and more slowly as the number of hops increases.

3.5.2 Diffusion and Conduction

3.5.2.1 Random Walks and Diffusion If one carefully places a drop of cream in a cup of coffee and if the coffee is not stirred, it will take a long time for the cream and coffee to mix. The cream is diffusing into the coffee, and diffusion is slow because it is described by a random walk.

A molecule of cream placed in the coffee moves rapidly because of its thermal energy. Its speed is comparable to the speed of sound. This suggests that cream should quickly spread throughout the coffee cup. However, every cream molecule can travel only a tiny intermolecular distance before it collides with another molecule. After the collision, it moves in a random direction. As with the frog journey, this means the path is a random walk, and the net displacement is proportional to the square root of the time.

$$\text{diffusion distance} \approx \sqrt{6D \times \text{time}}$$

The constant D in the equation is the "diffusion constant," defined in a way that necessitates the extra factor of 6. The diffusion constant is proportional to the product of the molecular speed and the distance traveled between collisions. A typical diffusion constant for water is $D = 2 \times 10^{-9}$ meter^2 / second. Using this number in the equation relating time and distance shows that diffusion of 1 centimeter should take about 15 minutes. However, the square-root dependence on the time means it would take a month for 1 meter of diffusion.

Does this really work for cream in coffee? In principle, the diffusion calculation is correct, but it is extremely difficult to completely eliminate the fluid currents, which stir the cream much more rapidly than diffusion. This is especially true for hot coffee because warmer coffee rises and replaces the cooler coffee at the surface.

Biology uses diffusion in many ways. Oxygen diffuses from the lung alveoli (very thin-walled chambers) into the blood, and carbon dioxide diffuses the opposite way. Many materials diffuse through cell walls. The slowness of diffusion over a long distance is one reason biological membranes are so thin.

3.5.2.2 Heat Conduction If the temperature on the outside of a window is freezing and it is room temperature on the inside, conduction tells you how rapidly heat flows out the window. Reducing heat conduction can significantly reduce the cost of heating homes and businesses.

Heat can be transferred from hot to cold in many ways. For window glass and many other solids, energy diffusion is the method of conduction. Atomic collisions typically transfer energy from the higher-energy (hotter) atom to the lower-energy (colder) atom. This diffusion is the random walk of extra energy through the forest of atoms that make up the solid. If a small spot in a solid is heated, the distance the heat diffuses away from the hot spot is described by the same diffusion formula.

$$\text{conduction distance} \approx \sqrt{6k \times \text{time}}$$

For heat conduction, the diffusion constant D is replaced by the "thermal diffusivity" κ. The range of thermal diffusivities is enormous because some materials are good heat conductors and others are nearly perfect insulators. Silver is an especially good thermal conductor, with $\kappa = 1.66 \times 10^{-4}\,\text{meter}^2/\text{second}$ because mobile electrons (instead of atoms) carry most of the energy. For quartz, κ is 100 times smaller and the thermal diffusivity of rubber is 1000 times smaller.

At teatime, it is easy to know if you are being served with the best dining utensils. One quickly notices the hot tea warming the handle of the silver spoon. Because the thermal diffusivity of pure silver is about 40 times larger than that of stainless steel, cheap imitation "silverware" is not so hot.

3.5.3 Age of Earth

Lord Kelvin used thermal conductivity to misestimate the age of Earth. He started with the equation that relates conduction distance, thermal diffusivity, and time. The goal was to solve for the time. Kelvin himself measured the diffusivity of rock and obtained a reasonable value: $\kappa = 0.9 \times 10^{-6}\,\text{meter}^2/\text{second}$. This was the easy part. Estimating the conduction distance and justifying use of the equation is much more difficult.

Kelvin assumed that Earth began when it changed from a molten ball to a solid sphere. He took the melting temperature to be that of rock, around 4000 K (a slight overestimate) and assumed Earth's interior remains near the melting temperature. He estimated the conduction distance by looking at temperature increases in deep mines. Observations suggested that every kilometer of depth is associated with a temperature increase of 35 K. If this temperature increase persisted to great depth, then dividing 3700 by 35 means the temperature approaches the melting temperature at about 100 kilometers, so 100 kilometers is a reasonable estimate for the conduction distance. Compared to 6400 kilometers (Earth's radius), the 100-kilometer conduction distance means only a thin outer shell of Earth has cooled, as shown in Figure 3.14.

The next step is algebra. Substituting the values for the conduction distance and the thermal diffusivity into the equation for the conduction distance gives

$$100 \text{ kilometers} = \sqrt{\frac{6 \times 0.9 \times 10^{-6} \text{meter}^2}{\text{second}} \times \text{time}}$$

This can be solved for the time by squaring both sides. Because there are about 31 million seconds in a year, the result is

Age of Earth = 60 million years (roughly)

Today, we know that Earth's age is 4.5 billion years, so the calculation is 75 times too small. This provides a cautionary tale about physics

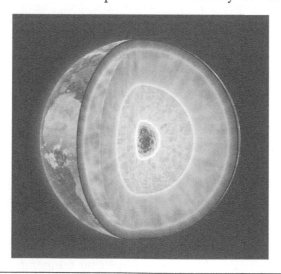

Figure 3.14 Cross section of Earth shows only a thin layer at the surface is cool.

calculations. Even when they are performed by the most skilled scientists, a mistaken initial assumption can lead to absurd results.

3.5.3.1 The Controversy In 1862, Kelvin published calculations that estimated Earth's age to be between 20 million and 400 million years. His estimate was not precise because of the uncertainties involved in his assumptions. His later results narrowed the uncertainty and pushed the result toward a younger age of 20 million years.

Just 3 years before Kelvin's first age calculation, Charles Darwin published *On the Origin of Species*. Although Darwin avoided age estimates in later versions of his book, it was clear that Kelvin and Darwin had reached very different conclusions on when the world began. Evolution takes a long time.

Darwin was not the first to see evidence of an ancient world. In 1830–1833, Charles Lyell published *Principles of Geology*. This book furthered the developing idea that geology acts slowly. It takes many millions of years to build a mountain. Charles Darwin carried *Principles of Geology* with him on his voyage of the *Beagle*. Despite Kelvin's calculations, geology and biology agreed that our Earth is very old earth.

Lord Kelvin (named William Thomson until 1892) was one of the greatest scientists of his time. Kelvin had enormous prestige, and he was not the least bit shy in insisting on the validity of his age results and in finding fault with those who disagreed. The consequence was a public controversy between scientific disciplines. Many took Kelvin's side, using reasoning similar to the following.

> Some of the great scientists, carefully ciphering the evidences furnished by geology, have arrived at the conviction that our world is prodigiously old, and they may be right but Lord Kelvin is not of their opinion. He takes the cautious, conservative view, in order to be on the safe side, and feels sure it is not so old as they think. As Lord Kelvin is the highest authority in science now living, I think we must yield to him and accept his views.

This quotation from Mark Twain reminds us that ideas should be judged on the soundness of the reasoning and experimental evidence. Using the reputation of a scientist as a judge of validity is a lazy shortcut that often fails.

Kelvin's mistake was not as simple as an algebraic fumble. It was his overconfident view that physics was essentially complete and his physical model was valid, so he could deal with a problem as complicated as the age of Earth.

The error can be traced to two sources. The first was the radioactivity observed by Henri Becquerel 1896. Because radioactivity was discovered 34 years after Kelvin's initial calculations, he cannot be blamed for ignoring this unknown phenomenon. Kelvin had reasoned that because Earth's interior is still hot, it could not have cooled for billions of years. Today, we know that a significant portion of Earth's heat is continuously generated by the radioactive decay of buried potassium, uranium, and thorium. Thus, Earth can be hot even though it is very old. Late in life, Kelvin reluctantly admitted problems associated with radioactivity, but there was never a published retraction.

Kelvin's second error was the assumption that Earth conducts heat as if it were a solid. At depths of less than 100 kilometers, the earth is not solid, and even slow fluid flow can carry heat from the center to the crust more efficiently than conduction. Kelvin must have known about volcanoes and molten rock, so he is not above criticism on this point. He was wrong for the same reason that cream rapidly mixes with coffee when the coffee is stirred, even if the stirring is very slow.

3.5.4 Brownian Motion and Equipartition

In 1827, biologist Robert Brown examined pollen grains in water through a microscope. He noticed that they hopped about in an unpredictable way. The motion was a random walk. Many years later, Albert Einstein and others explained this random walk in terms of atomic collisions. Even though Brown may not have been the first observer and even though he had no explanation for the random hopping, it is called *Brownian motion*.

The random motion of small objects suspended in a gas or a liquid is produced by collision with the fluid molecules. An understanding of Brownian motion requires a description of the acceleration produced by molecular collisions.

3.5.4.1 Speed Increase: Half the Story   A pollen grain is placed in helium gas. Each time a helium atom bumps the pollen, the collision gives

the pollen a "kick." The kicks are random, so the pollen speed undergoes a random walk. The random walk equations are still valid if one changes the hopping distance with the pollen grain speed. The result suggests that the pollen speed should increase with the square root of the time, just as diffusion causes displacement to increase with the square root of time. This does not make physical sense because pollen grains do not go increasingly faster as you watch them. There is a compensating mechanism that slows the pollen motion.

3.5.4.2 Speed Decrease: The Other Half   The pollen is slowed because its motion through the helium produces more collisions from the front than from behind. The faster the pollen moves, the more rapidly it will be slowed. The correction for the excess number of collisions from the front is proportional to the ratio of the pollen speed to the helium atom speed v.

3.5.4.3 Equipartition   The average speed of a pollen grain is determined by equating the *speed increase* produced by the helium atom kicks and the *speed decrease* resulting from the preferential head-on collisions. The result says the kinetic energy of the pollen grain and the helium atom are the same.

$$\frac{1}{2}MV^2 = \frac{1}{2}mv^2$$

This equality on average applies even though the mass M of the pollen grain is much larger than the mass m of the helium atom. Equal kinetic energies with very different masses mean the pollen is moving much more slowly than the helium atoms. A more methodical approach to thermal physics gives solid proof that these two kinetic energies are the same. This energy equality is an example of *equipartition*.

3.5.4.4 Equipartition Generalizations and Consequences   If a pollen grain (comparatively huge) has the same average kinetic energy as any one of the bombarding helium atoms, it is reasonable to surmise that objects of any size would have the same kinetic energy. This suggestion is correct. In particular, in a mixture of helium and argon atoms, the two types of atoms have the same average kinetic energy. That means the result established for helium,

$$\frac{3}{2}kT = average\ atom\ energy$$

applies to all simple gas atoms. Because argon atoms are 10 times more massive than helium atoms, their speed is slower by the factor $\sqrt{10} = 3.16$. Helium atoms are so fast that they occasionally escape from Earth's gravity. Escape is much less likely for the slower argon atoms. The escapes explain why helium in our atmosphere is more than 1000 times less common than argon, even though helium is a much more abundant element of the universe and our solar system.

The energy equality applies only to the kinetic energy associated with straight-line motion. So, if helium atoms are mixed with oxygen molecules, the average kinetic energies of straight-line motion are the same for the two gas components. Oxygen molecule rotations and vibrations add to the energy at high temperatures.

Equipartition extends to harmonic oscillators like the pendulum. Because a simple harmonic oscillator moves in one dimension, it is natural to associate $(1/2)kT$ with the average kinetic energy of this motion. But, a pendulum (and all harmonic oscillators) also has potential energy, and the average potential energy is equal to the average kinetic energy. This means the equipartition theorem leads to a simple-looking result.

Average thermal energy of all harmonic oscillators is kT.

A simple model of solid vibrations is based on harmonic oscillators. Each atom vibrates in a "cage" formed by the neighboring atoms. Because the atoms can vibrate in three dimensions, one obtains

Solid vibration energy = $3kT \times$ (Number of atoms)

Quantum mechanics explains why this result is only valid at high temperature.

A generalization of equipartition to the harmonic oscillators that form electromagnetic radiation leads to nonsense. If each mode of oscillation had a thermal energy kT, the energy would be infinite because the number of different waves is infinite. This absurd result is eliminated by quantum mechanics.

ELECTROMAGNETISM

4.1 Introduction

The forces of electromagnetism are the second half of classical physics. Newton's mechanics is the first half. Up to 1900, all of physics was ultimately based on Newton, gravity, and electromagnetism. Some were confident that the foundations of physics were established and only details remained. This era of confidence did not last long.

Electricity can appear to be quite different from magnetism, but the two are really part of a unified whole. This unity is expressed in Maxwell's equations, which are the culmination of classical electromagnetism. The ensuing discussion of electricity is followed by magnetism and the "big picture" of unified electromagnetism.

4.2 Electrostatics

Electromagnetism simplifies and becomes electrostatics when the charges are not moving. This case is considered first.

4.2.1 Atoms and Charge

Atoms are the building blocks of the elements, and they are made from nature's basic charges. A model proposed by William Rutherford in 1913 describes the classical structure. Light electrons fly around a heavy nucleus. The nucleus is composed of protons and neutrons, but Rutherford did not know about the nuclear components. Every proton has the same positive charge $+e$. Every electron has exactly the opposite charge $-e$. The electron was first "discovered" by J. J. Thomson, in part because he produced a better vacuum for observing cathode ray tubes. The slightly annoying negative choice for the electron charge

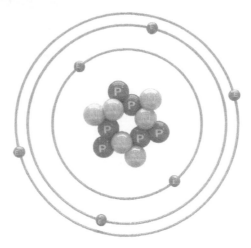

Figure 4.1 An idealized and distorted view of an uncharged carbon atom with six negative electrons, six positive protons, and six uncharged neutrons. In real atoms, the nucleus is much smaller, and quantum mechanics does not allow a precise determination of particle locations.

can be traced back to Benjamin Franklin. Neutrons have no charge. Even though the proton and neutron both have masses about 1800 times the mass of an electron, the central nucleus is 10,000 times smaller than the atom. An atom's dimension is determined by the distance the electrons stray from the nucleus. The nuclear size compared to the distance to the electrons is greatly expanded in Figure 4.1.

Normally, the electron and proton numbers (six each for the carbon atom of Figure 4.1) match up to make a neutral atom with no net charge. However, in some cases electrons can be removed from (or added to) atoms, giving objects a charge. If an object has 30 extra electrons, its charge Q is $-30e$. Today's standard unit of charge is the coulomb, analogous to the kilogram unit of mass. A coulomb is much larger than the charge of an electron, just as a kilogram is much larger than the mass of an electron.

4.2.2 Coulomb's Law

Coulomb's law gives the force F between charges Q and q separated by a distance r. It has the same geometrical underpinning as the universal law of gravity.

$$F = K \frac{Q \cdot q}{r^2}$$

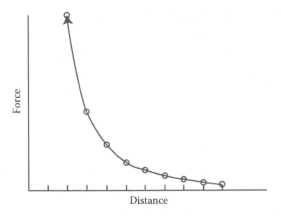

Figure 4.2 Equivalent curves describe the distance dependence of the forces of electrostatics and gravity because they both follow the inverse square law.

Both electrostatic and gravity forces become small for large separations because they decrease with the square of the distance r. The inverse square law is shown again in Figure 4.2.

The similarity between electrostatics and gravity can be overstated. The charges Q and q in Coulomb's law that replace the masses M and m in the law of gravity can be either positive or negative, but masses are always positive. The electrostatic force is attractive for charges of opposite sign and repulsive for charges of the same sign, but the gravitational force is always an attraction. The fundamental constant K determines the strength of electrostatics in the same way that the fundamental constant G determines the strength of gravity.

Pierre-Augustin de Coulomb invented and then used a torsion balance to measure forces between charges. Using this balance, he verified the inverse square relation around 1785. A few years later, Henry Cavendish used a more sensitive version of this torsion balance (see Figure 1.3 in Chapter 1) to measure the gravitational constant G.

Coulomb's law seems less evident than gravity, but there are many clear examples of the electrostatic force. The little pieces of Styrofoam called "packing peanuts" stick to surfaces because they are light and easily acquire a charge. Strands of dry hair can stick to a comb. Rubbing fur on plastic transfers electrons from the fur to the plastic. After this transfer, the negatively charged plastic attracts the positively charged fur.

On the atomic distance scale, the electrostatic attraction of protons and electrons explains why the atom does not fall apart. However, electrostatics does not answer two basic atomic physics questions. Why don't the negative electrons fall into the positive nucleus? Why doesn't the electrostatic repulsion of the protons cause the nucleus to explode? Quantum mechanics and an additional strong force are needed to answer these questions. They are discussed in Chapter 6.

When force is measured in newtons, the charge in coulombs, and distances in meters, experiments show that the constant K of electrostatics is very large. For example, if two 1-coulomb charges could be placed 1 meter apart, the repulsive force would be 9 billion newtons. This enormous force means 1 coulomb is much too large a charge to place on a small object. Because Coulomb's law works for any charge, a more physical example is the repulsion of two charges that are both only 10 microcoulombs. Then, the repulsive force at 1-meter separation is a reasonable 9/10 newton.

Ideally, Coulomb should have described the repulsion between two electrons because the electron holds the fundamental charge, but eighteenth-century physics (the age of Coulomb's experiments) was ignorant of electrons and protons. We have Robert Andrews Millikan (with help from others) to thank for the first accurate measurement of the electron charge about 125 years after Coulomb's work. Before looking at this famous experiment, definitions and some geometry are needed.

4.2.3 Electric Field

Coulomb's law can be viewed as an expression for an electric field, just as the universal law of gravity determines a gravitational field. With this viewpoint, the charge Q produces the electric field that extends throughout space. The electric field of Q is shown in Figure 4.3. When a second charge q is placed in the field produced by Q, it experiences a force. This is another example of structure in empty space. The potential energy of electrostatics is stored in the electric field of the vacuum. The electric field, being proportional to a force, has both a magnitude and direction and is denoted by the symbol $\vec{E}$. The force on any charge q placed in this field is

$$\vec{F} = q\vec{E}$$

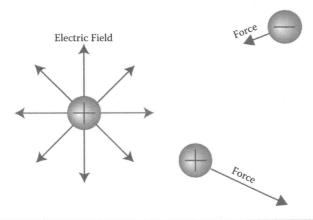

Figure 4.3 The electric field produced by the charge on the left extends to the other two charges. The force is parallel to the field for the positive charge and opposite the field for the negative charge. The force is smaller for larger separation. The two charges on the right also produce electric fields and corresponding forces, but they are not shown.

Because charges can be either positive or negative, forces can be either parallel or antiparallel to the electric field.

4.2.4 Geometry, Forces, and Fields

Geometry, in particular the $(4\pi r^2)$ surface area of a sphere, is the source of the inverse square law. For gravity, the spherical geometry of planets, moons, and stars is commonly encountered. Spheres are neither required nor useful for many electrostatic examples, and this means the inverse square law may be hidden. An important example is provided by two large parallel flat surfaces, as shown in Figure 4.4. One surface is uniformly covered with positive charge, and the other

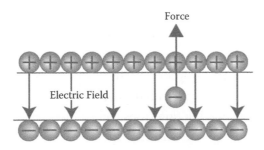

Figure 4.4 Positive and negative charges arranged on parallel planes produce an electric field pointing from the positive to the negative plane. The field exerts a force on the extra charge introduced between the planes.

is uniformly covered with an equal amount of negative charge. In the region between the two planes, the electric field is constant and proportional to the concentration of charge on the two planes. The magnitude of the field E can be measured by a determination of the force on a charge q placed between the planes and the formula $F = qE$.

4.2.5 Millikan Oil Drop Experiment

The famous oil drop experiment led by Robert Millikan determined the charge of a single electron. The electron's very small charge meant the measurement was a feat that required both ingenuity and patience.

The geometry of Figure 4.4 was used to establish an electric field. In principle, measuring the charge of the electron (called $-e$) should be simple. Just place an electron in the field and measure the force. Using $\vec{F} = -e\vec{E}$ and knowing both the field and the force, it is easy to solve for the charge $-e$. This is easier said than done. One cannot simply "place" an electron somewhere and measure the force on it. The electron must be attached to something and the force on that something must somehow be deduced. The tricks used to do this are described next.

The charged planes were placed one above the other with the positive charge on the top. A spray of tiny oil drops was drifted into the region between the charged planes. By chance (or by some coaxing), some of the oil drops had one more electron than proton, giving a charge $-e$ to the oil drop, as shown in Figure 4.5. There were then two forces. Gravity pulled the drop down, and the electric field pushed up on the attached electron. When electrostatics and gravity exactly canceled, the drop neither fell nor rose. The equality of forces means $eE = mg$. Thus, knowing the mass m, g, and the electric field E determined the electron charge.

One trick remained. The mass of a tiny oil drop is not easy to measure. This difficult measurement was done by removing the electric field and allowing the drop to fall. The bigger and heavier the drop, the faster it falls. This is a familiar phenomenon. Fog droplets take a long time to settle, but big raindrops come down pretty fast. The falling drop quickly acquires "terminal speed" when the retarding drag force cancels gravity. Newton's simple expression for drag force is not accurate for the oil drops, but an improved formula accurately relates drag to drop size and determines the drop's mass.

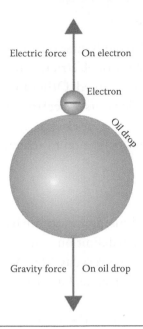

Figure 4.5 A properly adjusted electric field means the upward electric force on the electron cancels gravity's pull on the oil drop.

Finally, the job was done. The result was $e = -1.6 \times 10^{-19}$ coulombs. Knowing the electric charge, the electrostatic attraction of an electron to a proton can be calculated using Coulomb's law. On the scale of everyday life, this attraction is very small. If an electron and proton are separated by only 0.1 nanometer (a typical separation inside an atom), the force is only about 23 billionths of a newton, but this attraction holds all atoms together, and it is essential for the structure of everything.

The electron and proton also attract each other by gravity. However, the electrostatic electron-proton attraction is 10^{39} times larger than the gravitational pull. The number 10^{39} is so big (39 zeros after the decimal point) that, for all practical purposes, gravity plays no role in the properties of an atom.

It seems strange that the ratio of the two types of forces between fundamental particles should be such a large number. Some have speculated that this absurdly large number

1,000,000,000,000,000,000,000,000,000,000,000,000,000

must indicate something important. What is this relation between the two most fundamental forces trying to tell us?

4.3 Electricity in Practice

The electricity delivered to our homes makes modern life comfortable. The complicated technology of electricity that powers lightbulbs and television is often taken for granted. Other forms of electricity, including lightning and that which causes electrocution, are less comforting. Helpful or scary, most aspects of electricity are ultimately described by Coulomb's law and its application to materials.

4.3.1 An Analogy with Plumbing

Coulombs, volts, amps, ohms, and watts are basic quantities of applied electricity. But, their formal definitions can seem befuddling. This can obscure their importance in everyday life. An oversimplified plumbing analogy can be useful for some, but insulting to others—apologies to the others.

Someone has tied a knot in the water faucet of Figure 4.6 that limits the output. Two things determine how much water comes out. The first is the water pressure. More pressure delivers more water. The second is the knot. A tighter knot reduces the flow. These observations can be expressed (imprecisely) as an equation:

$$flow = \frac{pressure}{knot\ tightness}$$

Figure 4.6 How much water will flow from the knotted faucet?

In this equation, flow is the number of cubic meters of water that come through the faucet each second. Pressure is the force on each square meter of water before it goes through the knot. The water loses its pressure as it flows through the knot.

Pushing water through the knot takes work, so energy must be supplied by the pump producing the pressure. The work done is the product of the pressure times the volume of water. An equation summarizes this:

$$Energy = (Water\ volume) \times (Pressure)$$

The rate of doing work is the power. It is obtained by dividing the work by the time. Because flow is the water volume per unit time, the result is

$$Power = (flow) \times (pressure)$$

The electric circuit analogous to the knotted faucet is shown in Figure 4.7. The battery plays the role of the pump. It supplies the voltage (i.e., pressure) and the energy that drives the electric current through the wires. Electric current is measured in coulombs per second or amperes. This current is analogous to the fluid flow through the faucet. The lightbulb is equivalent to the knot. Its resistance slows the current and uses up the power supplied by the battery. The resistance is measured in ohms. The cumbersome words of electricity were chosen to honor

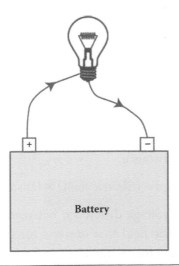

Figure 4.7   A basic circuit with similarities to the plumbing example.

Table 4.1 Plumbing and Electric Circuit Terminology

PLUMBING	ELECTRICITY	ELECTRIC UNITS
Flow	Electric current	Coulombs/second or amperes
Pressure	Voltage	Volts
Knot tightness	Resistance	Ohms
Power	Power	Watts

Charles-Augustin de Coulomb (1736–1806), Andre-Marie Ampère (1775–1836), Alessandro Volta (1745–1827), Georg Ohm (1789–1854), James Prescott Joule (1818–1889), and James Watt (1736–1819).

One can replace the equations describing plumbing with electric circuit equations by changing the terminology, as described in Table 4.1.

Using the conversions in Table 4.1, the plumbing equations become the equations for electricity. The relation between current, voltage, and resistance is called Ohm's law.

$$current = \frac{voltage}{resistance}$$

Section 4.3.4 comments on the development of Ohm's law.

The energy is the product of charge and voltage.

$$Energy = (Charge) \times (Voltage)$$

(A useful atomic-scale measure of energy is the electron volt [eV] or the energy change of 1 electron when the electric potential is changed by 1 volt: $1 eV = 1.6 \times 10^{-19}\ joules$.)

The power is the product of the electric current and the voltage:

$$Power = (Current) \times (Voltage)$$

4.3.2 Voltage and Electric Field

The relation between voltage (energy/charge) and electric field (force/charge) is the same as the relation between energy and force. Because energy is force times distance,

$$Voltage = (Electric\ field) \times (Distance)$$

Whenever there is a voltage difference between two points, there is a corresponding electric field that exerts a force on electrons. If two objects have different voltages, the electric field becomes larger as the objects are brought near each other.

4.3.3 Conductors and Insulators

The electric current in Figure 4.2 moves along metal wires. The wires are conductors because electrons can move easily through metals. The current cannot flow through the air because it is an insulator. The filament in the lightbulb is a poor conductor because it is a thin wire. Its resistance is large, and most of the battery's voltage is used to push the electrons through the filament's narrow channel. As they pass, the electrons bump into atoms, and their energy is transferred into various atomic excitations. This heats the filament to such a high temperature that it produces electromagnetic radiation (it glows). The filament is enclosed in glass and isolated from oxygen so it will not burn up.

4.3.3.1 A Numerical Example

Assume 6 coulombs pass through the bulb in 2 seconds. That means the current is 3 amps (coulombs per second). Assume the battery voltage is 6 volts. The power is the product of the 3 amps and the 6 volts, or 18 watts. The resistance is 6 volts divided by two amps, or 2 ohms.

4.3.3.2 Variations in Resistance

Every object has its own resistance to current. An iron wire has six times the electrical resistance of a copper wire of the same dimensions. This is a good reason to use copper when wiring a house. Silver has even less resistance than copper, but it costs a thousand times more. Wiring a house with silver would be silly.

The resistance of an insulator like glass or air can be 10^{20} times as high as the resistance of copper. These 20 factors of 10 mean essentially no electrical current can flow. The electrons in insulators are immobilized because they are bound to the atoms. However, if the electric field is large, it can sometimes detach atomic electrons and produce a bit of current where none should be seen.

4.3.4 Ohm's Law

The relation between current, voltage, and resistance (Current = Voltage/Resistance) is called Ohm's law. Sometimes sensible scientific ideas take some time to gain acceptance. When Georg Ohm published his law in

1827 (with slight modifications), it met with ferocious criticism. For example, a German minister of education said:

A physicist who professed such heresies was unworthy to teach science.

This probably says more about ministers of education than it does about Ohm.

Although the criticism of Ohm was mostly unjustified, one should not elevate Ohm's law to the same status as the laws of Newton or Coulomb. There are cases for which Ohm's law does not apply. The electrostatic spark and a lightning bolt are two examples of a large electric field that can liberate electrons. Other exceptions to Ohm's law are the basis of various modern electronic devices.

4.3.5 Charge Storage and Memory: The Capacitor

The object shown in Figure 4.8 is a "capacitor" because it has the capacity to store equal and opposite charges on the two flat surfaces. The charges placed on the conducting surfaces are kept separate by an insulating gap. The constant electric field between the two conducting surfaces means the voltage difference between the surfaces is the product of the electric field E and the width of the gap d.

For a fixed voltage, the stored charges are proportional surface area multiplied by the electric field. Large amounts of charge are stored by maximizing the surface area and minimizing the separation. (Millikan used a capacitor in his oil drop measurement of the electron's charge; see Section 4.2.5).

The early days of electricity research used primitive bulky capacitors called Leyden jars (invented in 1745) to store charge. An example is shown in Figure 4.9. Typically, the charge was produced by rubbing a material such as amber. Repeatedly touching charged amber to the metal connected to the jar's interior allowed the storage of fairly large

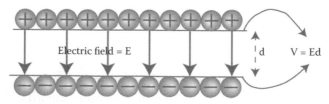

Electric field = E d $V = Ed$

Figure 4.8 A capacitor stores charge.

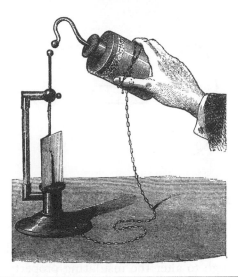

Figure 4.9 Leyden jar capacitor. A spark can appear when electrons hop between the jar and the stand. (*Trousset Encyclopedia*, 1886–1891.)

charges. An equal and opposite charge was attracted to the outside of the jar that was connected to the ground. People learned that one of these jars could store more charge when the inside and outside were covered with metal foil and the thickness of the glass separating the foils was reduced. A "battery" was originally an array of Leyden jars connected together to produce a larger charge.

Miniature capacitors serve as memory elements for computers and other electronic devices. The number 1 corresponds to a charged capacitor, and 0 means the charge is gone. Capacitors are the "pixels" of digital cameras, with the amount of charge deposited determined by the light intensity. Cell phones, televisions, and many other electronic devices depend on the charge storage ability of a huge number of miniature capacitors.

Capacitors are also part of electronic versions of harmonic oscillators, as described in Section 4.5. Their stored charge is a form of potential energy analogous to the height of a pendulum. Electronic oscillators are used to generate and receive electromagnetic waves.

4.3.6 Electrostatic Spark

A person can rub his or her feet on a rug and then touch a doorknob. The resulting spark (see Figure 4.10) can be surprising but not really

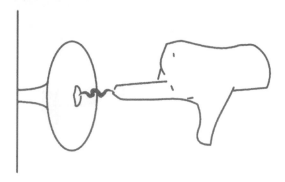

Figure 4.10 An electrostatic spark.

painful. Electric charge jumps the gap between finger and doorknob even though air is an insulator. The jump occurs when the electric field is large enough to alter the insulating property of the air. The process can start with just one free electron. Ordinarily, this electron would produce negligible electric current. However, a strong electric field can accelerate an electron to a high speed (and high kinetic energy). Then, a collision with an air molecule can detach a second electron from the target molecule. Repeating this process leads to a chain reaction, with an avalanche of freed electrons increasing rapidly. This "electric breakdown" can quickly change insulating air into thin filaments of conducting plasma (electrons separated from nuclei) where many free electrons are mixed with the air.

A rough estimate of the electric field needed to make electrons move through the air is 100,000 volts/meter. Assume an electrostatic spark jumps across a 3-millimeter gap (less than an eighth of an inch) between finger and doorknob. Using $E = 100,000$ *volts/meter,* $d = 0.003$ *meters,* and the equation $V = E \cdot d$ means there is a 300-volt difference between finger and doorknob.

The 300 excess volts are produced by the charge that was picked up by rubbing feet on the floor. Only a small charge, around 0.1 microcoulomb, is sufficient to make this much voltage. (Again, this is a rough estimate.) When the spark occurs, the excess charge is allowed to escape, and this lowers the energy. The energy change is the product of the 0.1-microcoulomb charge and 300 volts. This energy is only 30 microjoules. The released energy becomes heat with a little light and perhaps a little sound. The light is the spark, and the heat produced at the tip of the finger is part of the shock sensation.

4.3.7 *Lightning*

The lightning strike of Figure 4.11 is a gigantic electric spark. A process analogous to rubbing feet on a carpet moves a negative electric charge to the bottom of a cloud and a positive charge to the top. The accumulated charges are so large that the electric breakdown can be several kilometers long. This lightning strike may be between the top and bottom of the cloud. This "cloud lightning" can be seen from large distances, and it can be dangerous for airplanes. Lightning is usually more serious when the negative charge at the bottom of the cloud jumps to prominent objects on the ground, such as Lee Trevino's golf club. Lightning's transported charge, energy, and danger are enormous compared to the spark between finger and doorknob.

A small electrostatic spark generates a little light and some heat. Multiply the light and heat many times over to appreciate the light and heat of a lightning bolt. The heat causes the air to expand quickly, resulting in a loud noise (thunder).

Lightning is constantly delivering negative charge to Earth's surface. This results in a negatively charged Earth and an electric field that repels negative electric charge. This field results in a flow of electrons away from Earth that matches the charge delivered by lightning strikes.

This description of lightning is vague because there is much yet to learn. Physics asks for numbers, not superlatives. One would like to know how much charge, how much voltage, and how much energy

Figure 4.11   Lightning.

are associated with lightning. One would also like to know how the positive and negative charges are separated. There are many kinds of lightning, and reliable measurements are difficult, despite the early success of Benjamin Franklin, his son William, their kite, and their key. Franklin was lucky not to have killed himself in his lightning-kite experiment. (He was also lucky to have escaped serious injury in his attempt to kill a turkey with electricity.)

Ball lightning is the most mysterious. This lightning-like glowing blob suspended in space is so strange and seemingly unphysical that some people have questioned its existence. The ball can be up to meters across and can last a few seconds. The most likely explanation says the ball is made of glowing material blasted from the ground by an ordinary lightning strike.

4.3.8 Batteries

Batteries store chemical energy and change it into electrical energy. It is easy to make a mediocre battery. Iron and copper nails inserted into a potato or a lemon will produce some voltage. Despite the claims of some, the 2000-year-old "Baghdad battery" was not a battery, and its existence is not an indication that advanced aliens gave lessons on electricity and electroplating.

Alessandro Volta demonstrated the first battery in 1800, but a basic understanding had to wait 30 years for Michael Faraday's insight. Modern batteries are not nearly as simple as Figure 4.12 suggests, but the idealization illustrates the basic physics.

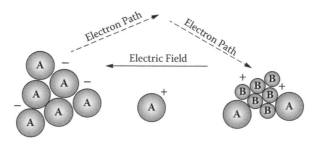

Figure 4.12 A simplified battery. A+ ions move to the B atoms, leaving extra electrons behind and bringing extra positive charge to the B atoms. Giving the electrons a conducting path from A to B can light a bulb.

The negative terminal of the simplified battery consists of atoms (call them A). The positive terminal is made up of different atoms (call them B). The chemical energy of the A atoms would be lowered if they could combine with the B atoms. However, there is a problem. The material in the space between the A and B atoms is an "electrolyte." The electrolyte has three important properties: (1) It is an insulator for electrons, so they cannot pass from A to B. (2) It does not allow neutral atoms to pass from A to B. (3) It allows A^+ ions to pass to the B terminal. An A^+ ion is positively charged because it has lost an electron.

When a battery is first assembled, a few of the A^+ ions move over to the lower-energy environment of the B atoms. However, because they can only flow as ions, a positive charge is quickly built up at the B terminal. The electrons abandoned by the A atoms are left at the A terminal. The charges on the terminals produce an electric field in the electrolyte that opposes the motion of additional A^+ ions. The flow stops when the electric potential energy (voltage times charge) cancels out the lower chemical energy of the paired A and B atoms. If the energy obtained by combining the A and B atoms is 1.5 electron volts, then the voltage difference between the battery terminals will be 1.5 volts.

Connecting the battery terminals to a lightbulb allows the electrons to flow from A to B. In the process, they give up their energy in the form of light and heat. As soon as the electrons start to flow, the voltage difference between the battery terminals is reduced. This allows more A^+ ions to join the B atoms, and the battery supplies additional charge to power the lightbulb. A battery is much different from an electric spark or a lightning strike because it can supply a nearly constant voltage even when there is an electric current. The battery is discharged only after all the A and B atoms are combined.

Most batteries involve much more complicated chemistry than this simple example. In many cases, the electrolyte plays an active role and participates in a series of chemical reactions that produce the battery voltage.

4.3.9 Car Batteries

Car batteries with 12 volts can supply hundreds of amps to start a car, corresponding to thousands of watts of power, but this power can be delivered only until the battery's chemical warehouse of energy is exhausted. Under ideal circumstances, each kilogram of a car's lead-acid battery can supply nearly 146,000 joules of energy before it is depleted. If all this energy could be changed to mechanical energy with 100% efficiency, it would be enough to lift the battery about 15 kilometers into the air. Despite this impressive-sounding energy storage, batteries are not that great. Burning a kilogram of gasoline supplies more than 300 times the energy stored in 1 kilogram of lead-acid battery. Modern batteries based on lithium can store 10 times the energy of the lead-acid battery, but they are still poor energy storage competitors of petroleum.

Electric cars have no need for an internal combustion engine or a gas station. Batteries supply the energy to electric motors. One should not assume that this battery power is the key to nonpolluting energy. A battery only stores energy. When battery energy is drained, it must be recharged. If the replacement energy comes from a coal-fired power plant, little is gained. Electric cars would be much more practical if reliable and long-lived batteries could be developed with better energy storage capacity. Battery development is an active area of research, but the problems are difficult. Great leaps forward in battery technology may occur soon, but they may not. It is as difficult to project scientific breakthroughs as it is to predict horse races. Doubled research money does not ensure doubled progress.

4.3.10 Electricity in the Home

For most people, electricity is delivered through conducting wires from an electric power plant. The voltage for household power is not constant like a battery voltage. It oscillates in time, as is shown in the sine wave of Figure 4.13. Typically, the frequency is 60 hertz, and the voltage varies

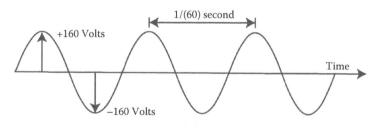

Figure 4.13 Time dependence of the voltage supplied to homes by power plants.

from about +160 volts at the positive peak to –160 volts at the negative trough. Appropriately averaged, this amounts to about 110 volts.

Some household appliances, such as electric clocks, take advantage of the 60-hertz oscillations. For other applications, like a lightbulb, the time variation of the voltage is not important. As with a battery, one must connect a wire between the two sides of an electrical outlet to produce a current. This completed circuit is needed so electrons have a place to go. The oscillating voltage means the current will oscillate similarly, going back and forth through the wires. This explains the term *alternating current* (AC). The constant voltage of a battery produces *direct current* (DC). Just as with the battery, the power delivered to the wire is equal to the product of the current and the voltage.

A toaster, hot plate, electric oven, and a space heater are all basically a wire connecting the two sides of an electrical outlet. The power from the electricity heats the wire. Because heat flows from hot to cold, the hot wire warms its surroundings. Soon, for instance, you have toast.

4.3.11 Electricity and Danger

Electricity can kill. The danger is not simply the electrical energy that is changed to heat. A human heart depends on small electrical signals to keep a steady rhythm. External electricity can interfere with these signals, and this can be bad news for the heart. A current of more than 0.1 amp through the chest is potentially lethal (Figure 4.14).

Although current, not voltage, characterizes danger, current is produced by a voltage difference. Current flows only when there is a completed circuit. That means if you are exposed to a high voltage, there must be a path for the electricity to flow through you. This is easy to demonstrate. Touch your tongue to one terminal of an ordinary battery and you will notice nothing. Touching both terminals with your tongue completes the circuit and allows the electric current to pass through your tongue. This can be unpleasant, but not lethal even for a high-voltage battery because no electricity passes near the heart.

One might think that voltage alone should be the correct measure of danger. After all, Ohm's law says the current is proportional to the

Figure 4.14 An unwise maneuver.

voltage. More voltage means more current, and more current is definitely more dangerous. However, it is hard to find the current using Ohm's law because the resistance of a human body is hard to determine. Pure water is an insulator, but blood and other body fluids are salty. Electrical currents can flow through your body as ionic motion in fluids. Dry skin can impede the current flow because it is a good insulator. However, skin dampened by salty perspiration is much less effective in stopping the current.

There are records of people being killed by 110 volts from a home plug. There are also records of people who survived lightning strikes and people who have not died after an initial 1000-volt shock from an electric chair. There is no hard-and-fast rule about how much voltage is lethal, but there is a common-sense rule: **Be careful**.

4.4 Magnetism

Magnetism has often been associated with magic. This is not surprising. Many real magnetic phenomena appear almost magical. They include bird navigation, bacteria orientation, the aurora, cyclotrons, train levitation, compasses, and sunspots. However, mystical magnetic healing and mesmerism are fantasies.

For something as complicated as magnetism, it helps to break it into pieces:

1. Magnetic fields exert forces on moving charges.
2. Moving charges produce magnetic fields.
3. Magnets are almost the same as moving charges, so magnetic fields exert forces on magnets and magnets produce magnetic fields.

4.4.1 Basic Example

Two charges (q and Q) provide the simplest example for the comparison of electricity and magnetism. When these charges are not moving, their mutual repulsion (or attraction) is described by electrostatics and Coulomb's law ($F = KQq / r^2$). When both charges are moving, there is an additional magnetic force. The mechanism is shown in Figure 4.15.

Because motion is fundamental to magnetism, it no longer suffices to picture charges as little spheres with plus or minus signs. Instead, a suggestive portrayal of magnetic effects imagines charges to be tiny screw-like (helical) objects with their orientations determined by the direction of their motion. Visualize a charge (helical shape) as embedded in a fluid. The motion of a helix through the fluid produces

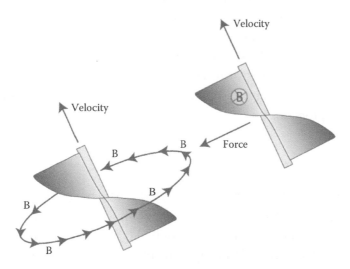

Figure 4.15 The moving charge at the lower left creates a circulating magnetic field. The second charge moving through this field is attracted toward the first charge.

circular stirring. The flow associated with the stirring is the magnetic field, denoted $\vec{B}$. If a second charge moves in this magnetic field, the flow from the first charge pushes sideways on the moving fins of the second charge. The cross with B in Figure 4.15 means the field is oriented into the page on the fins. The sideways force of attraction on the second moving charge is analogous to lift on an airplane or bird's wing. The second charge also produces a magnetic field (not shown to keep things simpler) that exerts the equal but oppositely directed attracting force on the first charge.

The geometry is complicated, but the result is simple. Two charges of the same sign moving parallel to each other, as shown in Figure 4.15, are subject to both the Coulomb force of repulsion and a magnetic attraction. Coulomb's force is always larger.

A magnetic field circling a moving charge seems more mysterious than the electric field that emanates from a charge. Indeed, when Hans Christian Oersted first noticed (during a class lecture in 1820) that a compass needle was deflected by an electric current, he initially assumed that the magnetic field radiated away from the current, but Oersted was no dummy. Within a few months, he figured out the correct circulating geometry of the magnetic field.

The directions of both the Coulomb and magnetic forces are reversed when the charges are of opposite sign. For either case, the magnetic force for charges moving together decreases the electrostatic force.

The diagram of Figure 4.15 is only a visual aid. Charges are not really helical, and there is no fluid to stir. However, the magnetic field and the forces between the charges are real. The magnetic force between moving charges seen in this simplest example reveals a fundamental problem.

4.4.2 Problem with the Basic Example: Relativity

This is the dilemma related to relativity: The extra magnetic force means an experiment performed on a moving train would not look the same to observers inside and outside the train. Magnetic fields and magnetic forces would be observed by people outside the train. However, for train passengers, the charges are not moving; there would be no magnetic field, no extra forces; only Coulomb's law should apply.

This paradox is even more perplexing when one notices that it is not clear what is moving and what is at rest. The earth is rotating and circling the sun. Also, the sun is moving around our galaxy, the Milky Way. Our galaxy is also moving. How can one calculate a force that depends on velocity when there is no absolute velocity?

The moving charge dilemma disturbed Albert Einstein. It is not surprising that Einstein's first paper on special relativity was titled "On the Electrodynamics of Moving Bodies." Einstein's special relativity eliminated the ambiguities of magnetism and returned physics to the solid foundation where physical observations are consistent for the person on the train and the person on the platform. For now, relativity is ignored because much of magnetism has an adequate non-relativistic description.

Relativity can often be ignored because it is frequently insignificant. For the example in Figure 4.15, the ratio of the magnetic attraction to the Coulomb repulsion is the tiny ratio $(v/c)^2$. Here, v is the speed of the charges, and c is the speed of light. A fast train travels at 100 meters/second, but the speed of light is 3 million times larger than that. Squaring the very small v/c gives an unmeasurably small magnetic correction to Coulomb's law.

Relativity is not always insignificant in the microscopic world. A typical speed of an electron in the hydrogen atom is $c/137$. The square of $1/137$ is not too small to be observed. Electron speeds in heavy atoms can be much closer to the speed of light. Relativity and magnetic effects are a necessity rather than a curiosity when describing a uranium atom.

4.4.3 Large Magnetic Fields

The small magnetic interaction of individual charges moving at ordinary speeds suggests that magnetism should be unimportant. That is clearly not the case. Magnetism can levitate trains. Magnetic fields become large when many charges move in unison. The example in Figure 4.16 depicts a swarm of circulating charges. The geometry means the magnetic field from each moving charge adds to produce a large field inside the loop. In practice, the charges can be electrons moving on a loop of metal wire. Coulomb repulsion of the moving electrons is not a problem because they are paired with an equal

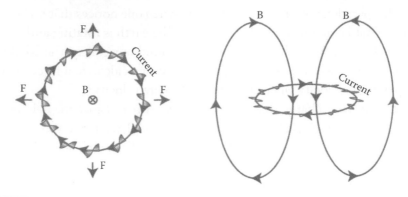

Figure 4.16 The combined magnetic fields of many circulating charges produce a large field. Left: The field at the center is directed into the page and denoted by the B and the cross. The field extends to the circle edge and pushes out on the current, as denoted by the force arrows. Right: A side view shows how the field circles the current loop.

number of positively charged (but stationary) ions in the wire. This means a large current of many amps (coulombs per second) can be easily produced even though enormous energy is required to confine a single-coulomb charge in a small space.

The magnetic field can be made much larger by wrapping the wire around the loop many times. Ten windings produce a magnetic field 10 times as strong. Nikola Tesla (see Section 4.5.2.7) used complicated electronics to develop very large oscillating magnetic and electric fields.

The tesla is the unit of magnetic fields. Earth's magnetic field is around 1/20,000 tesla. House magnets (refrigerator magnets) are stronger than this by a factor of 100 or more. The magnetic field of the Large Hadron Collider (LHC) operated by CERN (European Organization for Nuclear Research) on the Swiss-French border uses 8-tesla magnets. The largest magnetic field produced on Earth is about 45 tesla. Magnetic fields in exotic astronomical objects can dwarf all earthly magnetic fields. Observations suggest magnetic fields of "magnetars" are in excess of 100 million tesla. Strange physics would be the norm in these magnetic neutron stars.

4.4.4 Magnetic Energy and Pressure

Magnetic fields store energy. This invisible energy is analogous to the energy stored in an electric field. Both energies are proportional to the square of the field strength. Figure 4.16 shows the magnetic force pushing out on the current that is producing the magnetic field. This force can be associated with the pressure of the magnetic field that passes through the loop. Magnetic energies, forces, and pressures can be large. A powerful magnet with many windings and a large current can yield a force on the wires that is large enough to explode the magnet. Because energy is stored in the magnetic field, work must be done to create the current sustaining the field. The required work is described by Faraday's law in Section 4.5.2.

4.4.5 Motion in a Magnetic Field

The magnetic force depicted in Figure 4.15 is perpendicular to both the magnetic field and the direction the charge is moving. This peculiar force direction neither accelerates nor slows the particle. Instead, it guides charged particles into circular paths aligned perpendicular to the field direction. The charge can also maintain straight-line motion parallel to the field because there is no force parallel to the field. The combination of the circular and straight-line motion is a helical path like that shown in Figure 4.17.

The helical path means charges can move large distances parallel to the magnetic field, but their restricted motion perpendicular to the field results in "magnetic trapping." Van Allen radiation belts and controlled fusion experiments are two of many examples of magnetic trapping.

4.4.5.1 Van Allen Radiation Belts
Earth's magnetism extends beyond the atmosphere. In the nearly empty region thousands of kilometers above us, charges are trapped in "van Allen radiation belts" shown in

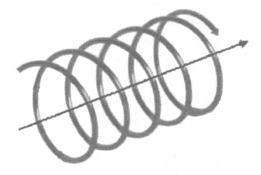

Figure 4.17   Path of a charged particle in a magnetic field. The field direction is indicated by the arrow.

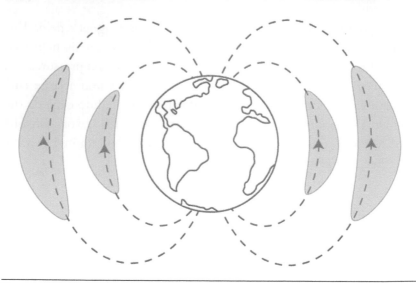

Figure 4.18 The Earth's magnetic field traps charged particles in helical orbits moving along magnetic field lines. The scale is distorted. The radiation belts really extend much further out. They are also distorted by the sun's solar wind.

Figure 4.18. These charged particles follow helical paths along the magnetic field. They sometimes enter the atmosphere, usually near the north and south magnetic poles. Interaction with the air transfers their energies into the light of auroras (northern lights and southern lights) as is shown in Figure 4.19.

Stronger magnetic fields near Earth's poles reflect many of the charges and drive them away from the poles. As a result, many charges stay trapped in these Van Allen radiation belts for a long time. The outer belt is mostly electrons, and the inner belt is electrons plus protons. The radiation belts extend far into space, and the flow of particles from the sun (the solar wind) distorts these radiation belts, so Figure 4.18 is an idealization.

Earth's magnetic field and the resulting radiation belts are not just a curiosity. Charged particles, mostly from the sun, are constantly bombarding us. Without our magnetic field acting as a shield, this particle flux would be lethal. One of many reasons advanced life is not possible on Mars is the lack of a protective Martian magnetic field.

4.4.5.2 Controlled Fusion The energy from our sun comes from nuclear fusion. This is a nuclear reaction that changes hydrogen nuclei to helium nuclei. In the process, a great deal of energy is released.

Figure 4.19 When charged particles trapped by Earth's magnetic field hit the upper atmosphere, their energy is transformed into impressive light displays.

Fusion can occur only at very high temperatures so that the nuclei collide with high speed. A fusion reaction related to the sun's fusion could be produced on Earth if the ions were sufficiently heated and compressed. However, the required temperature is so high that all containers would vaporize. Magnetically confining charged particles in a "donut" shape seems like a good way to achieve controlled fusion. The charges moving along the toroidal magnetic field should never find an escape route. Unfortunately, a variety of challenging technical problems means this "magnetic confinement" has never produced usable fusion energy. Scattering that ejects particles from the field is one of the problems. Unlimited energy from controlled fusion has remained a dream for more than 50 years. The geometry shown in Figure 4.20 only hints at the complexity of present-day controlled fusion experiments. Fusion is described in Chapter 6.

Fusion reactions take place in a hydrogen bomb, but this is not "controlled" fusion, and the enormous energy produced cannot be changed to useful work.

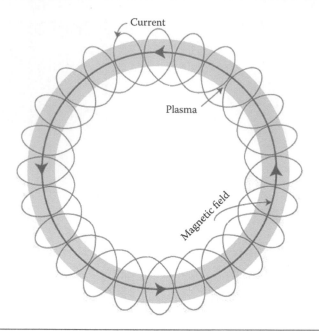

Figure 4.20 A donut-shaped magnetic field is produced by an electric current helix. Charged particles are trapped by the field for relatively long times. Real devices designed to produce controlled nuclear fusion are much more complex.

4.4.6 Orbit Period

Circular motion is an especially important special case of helical motion in a magnetic field. The magnetic force pulls the charge toward the center of the circle. The force is proportional to the charge Q, the velocity v, and the magnetic field B. Its magnitude is $F = QvB$. Combining this magnetic force formula with the formula for the acceleration of circular motion ($a = v^2 / r$) gives a surprisingly simple and enormously important result: The time for the charged particle to make one circular orbit (circumference divided by speed, or $2\pi r/v$) does not depend on the speed of the particle. Some algebra gives this time, called the *period*:

$$\text{Period} = \frac{2\pi}{B} \times \left(\frac{\text{mass}}{\text{charge}} \right)$$

Period and *frequency* are frequently encountered terms. The frequency is the inverse of the period, so a period of 1/10 second corresponds to a frequency of 10 hertz (or cycles per second). For motion in a magnetic field, the frequency is called the *cyclotron frequency*. The cyclotron frequency and its generalization appear in many areas of physics. Some examples follow.

4.4.6.1 Charge-to-Mass Ratio If one measures the period and the magnetic field *B*, the equation just discussed means one also knows the mass divided by the charge. The first to measure the mass-to-charge ratio of the electron using a technique related to this equation was J.J. Thomson. Thomson showed that the electron mass was a small fraction of the mass of any atom. Combining the Millikan oil drop experiment (reported in 1913) that accurately determined the electron charge with the mass-to-charge ratio (reported in 1897) tells us that the proton mass is about 1836 times the electron mass.

4.4.6.2 Cyclotron The cyclotron combines a properly timed electric field with a magnetic field to produce high-energy particles. The key to the cyclotron's success is the constancy of its frequency. The particle circles in the cyclotron's magnetic field. Each time the particle crosses from the left side to the right side of the cyclotron, a voltage is applied to push the particle ahead. By the time the particle crosses in the other direction, the voltage is reversed to give it another forward kick. As the energy increases, the orbit radius becomes larger. After many more orbits than are shown in Figure 4.21, the highest-energy particle flies out into a field-free region, where it travels in a straight line.

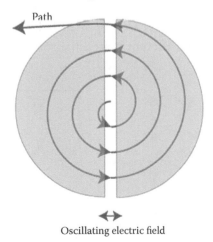

Path

Oscillating electric field

Figure 4.21 An idealized cyclotron showing the expanding orbit of a charged particle in a magnetic field. In real cyclotrons, many more orbits are needed to accelerate the particle.

One might think that timing the forward pulses of a cyclotron would be difficult, but it is surprisingly easy because the orbit period does not depend on the energy of the particle or the radius of the orbit.

There are limits to cyclotron energies. When the particle speed approaches the speed of light, there is a correction to the cyclotron frequency. Also, when charged particles travel in circles, they radiate electromagnetic waves and lose energy. These problems mean cyclotrons are not used to produce the highest-energy particles.

4.4.7 *Magnetic Materials*

Atoms are made of charges (electrons and protons). Magnetic fields exert forces on charges. As a result, when a material is placed in a magnetic field, the magnetic force sometimes leads to electric currents in the material. This is the source of magnetism. The total magnetic field in the material is the sum of the applied magnetic field and this "internal field." The magnitude and direction of the internal magnetic field depends on the material.

One can view the internal magnetism as the result of tiny electrical currents surrounding individual atoms. When the atomic electronic orbits are all aligned, the result is equivalent to a large electric current around the edge, as is shown in Figure 4.22. The resulting magnetism is analogous to the magnetism of the current loop of Figure 4.16.

For many materials called paramagnets, the internal field is only a small addition to the externally applied field. This correction can be important because different materials can be identified by their

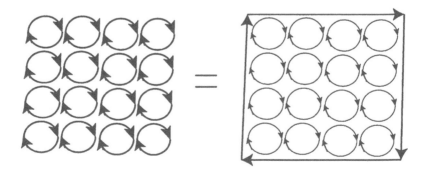

Figure 4.22 Atomic currents in some materials add together to make an average current surrounding the material. This current produces magnetism.

slightly different internal fields. In other cases, the internal field can be large. An extreme case is ferromagnetism because the internal magnetism persists even after the external field is eliminated. Iron, nickel, and cobalt are just three examples of many ferromagnetic materials.

4.4.8 Compass

The needle on a compass is magnetic and typically made of iron. Force on the compass needle can be determined using the equivalence between the magnetism of iron and the surface current shown in Figure 4.22.

When the compass needle is placed in a magnetic field, the force on the current exerts a torque on the needle that aligns it with the field. For simplicity, only a band of this current is shown on the cylindrically shaped compass needle in Figure 4.23. The aligned configuration shown on the left of Figure 4.23 corresponds to the lowest-energy orientation of the needle. Earth's small magnetic field is sufficient to align a properly suspended needle so it points toward "magnetic north."

In most places, magnetic north is fairly close to true north. But, there are differences that depend on location. The difference called

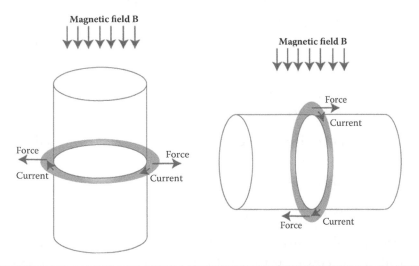

Figure 4.23 A magnetic field pushes on a loop of current until it is aligned as on the left. The alignment forces are shown on the right. When the loop represents a magnetized compass needle and the field is Earth's magnetism, this figure explains why compasses point north.

Figure 4.24 Compass north (magnetized needle) and true north (N) differ by about 20 degrees in Maine. The compass direction is closer to true north in most other parts of the United States.

magnetic declination can also change a little with time. A relatively large difference between true north and magnetic north is shown in Figure 4.24.

The extraordinarily talented Chinese scholar Shen Kuo (1031–1095) was the first person to note the difference between true north and magnetic north. A list of Shen's achievements is long and impressive, ranging from poetry to the observation of climate change. Shen's work is just one of many examples of the scientific accomplishments of China that have received scant attention.

4.4.9 Magnetic Moment

A magnet can be characterized by its energy when it is aligned with an external magnetic field. The term *magnetic moment* describes the energy. For alignment,

Magnetic energy = –(Magnetic field) × (Magnetic moment)

An especially simple magnetic moment is produced when a charge Q with mass m travels with speed v in a circle of radius r. An important calculation shows that the magnetic moment is proportional to the angular momentum of the circulating charge.

$$Magnetic\ moment = \frac{Q}{2m}(angular\ momentum)$$

This makes intuitive sense. Charge moving in a circle produces a magnetic field, and mass moving in a circle has angular momentum. To change angular momentum to magnetic moment, one divides out the mass and multiplies by the charge. The factor of 2 is a technicality.

Magnetic moment and magnetic field are very different. Earth's magnetic field is small, but it extends over such a large region that the magnetic moment is enormous. Magnetic fields produced by atoms can be large, but their size means the magnetic moment is small.

4.4.10 Force between Magnets

Magnets near each other, as shown in Figure 4.25, have lower energy when their fields are aligned end to end. The magnetic field of the upper magnet produced by a current loop extends to the lower magnet. The force of the current loop of the lower magnet twists it in the direction that makes the current loops parallel. Sometimes, one associates north and south poles with magnets. Then, the force that

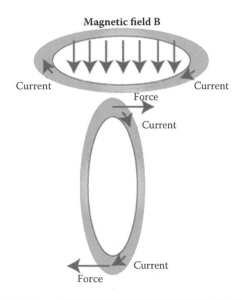

Figure 4.25 The upper magnet (represented by a current loop) produces a magnetic field that twists the lower magnet (another current loop) toward a parallel orientation. The energy is lowest when the magnets are aligned.

pushes them toward parallel orientation is equivalent to saying the North Pole attracts the South Pole.

Aligned magnets have lower energy, and the closer they are, the greater the energy savings is. The energy difference translates into a force. Aligned magnets attract. Similarly, magnets antialigned repel. The repulsive force can be large enough to lift heavy objects.

4.4.11 Electric Motors and Generators

An electric motor converts electrical energy into mechanical energy. Motor designs are complicated, but they all use the magnetic force on an electric current to produce motion. The current loop shown in the magnetic field of Figure 4.26 can rotate. The force perpendicular to both the current and the field will cause the loop to rotate until it is oriented with the field in a manner analogous to compass needle orientation.

To keep the loop rotating, the current in the coil is reversed just when alignment is achieved. The reversed current then pushes the loop another 180 degrees. Endless repetition of the reversals leads to rapid rotation and a conversion of electrical energy into mechanical energy. A simple, but clever, trick reverses the current at exactly the correct orientations to produce the continued rotation.

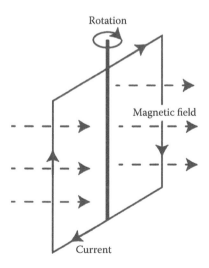

Figure 4.26 A loop of wire carrying a current experiences a torque that pushes its magnetic field toward alignment with the external magnetic field. If the current direction is reversed each time orientation is achieved, the loop will continue to rotate.

4.4.11.1 Electric Generator   A generator is an electric motor run in reverse, so mechanical energy is converted to electrical energy. Instead of allowing the magnetic force to push the coil of wire, the coil is rotated by a mechanical force, and the current is produced by the motion of the charges in the conducting wire past the magnetic field.

Just as the current must be periodically reversed to keep a motor running, the electricity generated by a rotating current loop will periodically reverse. Rotating the loop at 60 hertz produces the standard 60-cyle sine wave of the AC current delivered to homes.

4.4.12 Magnetic Bacteria   Compass navigation was invented by the Chinese about 1000 years ago, but primitive creatures only a millionth of a meter long had the idea first. A special bacteria type shown in Figure 4.27 contains small magnets. Earth's magnetic field aligns the bacteria just as it aligns compass needles, so their magnetic moments are parallel to the field. The bacteria tails propel them in the direction of the field. This is not done for bacterial amusement. It is a survival technique. These anaerobic bacteria are killed by excess oxygen, so it is good policy to swim toward deeper water. Earth's magnetic field has an up-down component, so if bacteria magnets are oriented to swim north, they will also swim to deeper water in the Northern Hemisphere. Reverse magnetic orientation is required for Australian magnetic bacteria, and magnetism will not help bacteria living near the equator. Bacteria with an accidentally reversed alignment, such as the upper example in Figure 4.27, die quickly before they can reproduce.

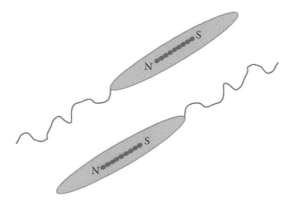

Figure 4.27   Magnetic bacteria with opposite magnetic cores. The magnets in the upper bacterium are improperly oriented. It will swim toward an oxygen-contaminated surface and die.

4.5 Unification of Electricity and Magnetism: Maxwell and Faraday

4.5.1 Maxwell's Equations

Maxwell's equations from the 1860s represent a monumental unifying achievement. Combined with the rules for the forces on a charge, they describe all of classical electromagnetism. Four sentences summarize Maxwell's mathematical description of electric and magnetic fields. The first two sentences relate to the geometry; the second two determine the time dependence.

1. *An electric charge produces an electric field radiating away from the charge.*
   Combined with the electric field's force on a charge, this summarizes electrostatics.

2. *An electric current (moving charge) produces a magnetic field circling the current.*
   Combined with the magnetic field's force on a moving charge, this summarizes magnetostatics.

A. *A magnetic field changing in time generates a circulating electric field. This field can lead to currents and magnetic fields that oppose the change of the original magnetic field.*
   This is essentially Faraday's law, which is described in the next section.

B. *An electric field changing in time generates a circulating magnetic field.*
   Maxwell's insight showed that this last statement was necessary for a consistent theory. Its validity proves that electromagnetic waves are composed of electric and magnetic fields.

Maxwell's equations are more than these four summaries of the concepts. They are a quantitative, exact, and complete description of classical electromagnetism.

4.5.2 Faraday's Law

The discovery of the electric field resulting from a changing magnetic field (sentence A) originated with the amazing physical insight of Michael Faraday. The amazing mathematical skills of James Clerk Maxwell filled out Faraday's ideas. Some applications of Faraday's law follow, but first a current loop example can keep things straight.

An electric current loop like the one in the top of Figure 4.28 produces a magnetic field.

> If the current and the magnetic field are increasing (decreasing), Faraday's electric field opposes the increase (decrease).
> If the current does not change, there is no electric field from Faraday's law.

4.5.2.1 Current Loop and Energy Conservation The magnetic field produced by a current loop contains energy. Faraday's law explains the origin of this energy and ensures energy conservation. As current in the loop is increased from zero, the changing magnetic field means Faraday's electric field opposes the current. A battery or some other voltage source is needed to overcome this opposing field. The battery supplies the energy stored in the magnetic field.

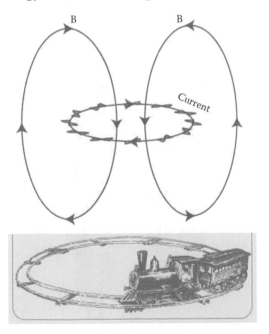

Figure 4.28 Faraday's law says a current loop carries energy and resists stopping, just as a speeding locomotive is difficult to stop. This analogy has its limitations.

4.5.2.2 Persistent Current A current loop is analogous (roughly) to a toy train on a circular track, as is suggested in Figure 4.28. The train has kinetic energy, and its momentum means it resists an attempt to slow it. Trying to suddenly stop the train by a barrier or break in the track would be difficult. The train might smash through the barrier or jump across the track. The same observation applies to the current loop. A sudden break in the wire carrying the current could result in a spark jumping across the break because Faraday's electric field tries to preserve the current.

The current-train analogy deserves skepticism. Current is uniformly distributed over the loop, while the train is only on one part of the track. The energy of the current loop is contained in the magnetic field; the train has ordinary kinetic energy and momentum. There is no easy analogue of the magnetic field for the train.

4.5.2.3 A Spark Plug A spark plug is a practical application of the persistence of current. When a switch changes the path of a current loop so it must flow through a spark plug, the current cannot stop suddenly, so it has no choice but to jump the spark plug gap shown on the right side of Figure 4.29.

Real spark plug circuits, even the ones for antique cars, are carefully engineered to maximize the spark. The single loop of wire is

Figure 4.29 The spark on the right-hand spark plug occurs across the small gap that can be seen on the left. The spark lasts only a short time after the current is deflected to the spark plug.

replaced by a coil with many windings. Also, the coil is wound around a piece of iron. Breaking the circuit eliminates both the external magnetic field of the coil and the internal magnetic field in the iron. This greatly increases the changing magnetic field, the resulting electric field of Faraday's law, and the ability of the system to make a spark.

4.5.2.4 Transformer
Transformers make the delivery of home electrical power much more efficient. The voltage in power lines is typically more than 100 times the house voltage. The power lines employ high voltage because this requires smaller current to deliver the needed power. Smaller electrical current means less energy is lost in the transmission. However, high voltage is dangerous, so it is reduced by a transformer before it is allowed into a home. The description that follows simplifies transformer physics.

Faraday's law means nature tries to avoid changing magnetic fields. As an illustration, assume an increasing electric current in one wire loop produces an increasing magnetic field. If there is another wire loop nearby that can carry a current, the Faraday electric field will produce a current in the second wire, as is shown in Figure 4.30. This second current will produce a magnetic field that cancels (or nearly cancels) the field of the first wire. The end result is a transfer of the current between the two wires even though they are not in contact. In the example in Figure 4.30, the solid wire is a double loop, so the current in the dotted wire must be doubled to cancel the magnetic field from the solid wire. This geometry "transforms" a current in the solid wire into double the current in the dotted wire.

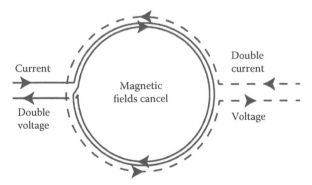

Figure 4.30 An idealized transformer. The currents in the solid and dotted wires are in opposite directions, yielding a nearly vanishing magnetic field.

Conservation of energy applies to a transformer. Thus, the power added to one side of a transformer must be extracted from the other side. Because power is the product of current and voltage, doubling the current cuts the voltage in half. Changing the number of times the solid and dotted wires surround the loop produces different transformation ratios of the current and voltage. Real transformers use many turns of wire and fill the loop with a magnetic material such as iron to make the transformation more efficient.

Faraday's law relates changing currents and magnetic fields. Transformers work only because power lines provide AC. A transformer cannot change the voltage of a battery because its voltage is constant.

4.5.2.5 Electromagnetic Harmonic Oscillator An electromagnetic harmonic oscillator is analogous to a pendulum. A pendulum moves back and forth as it exchanges kinetic energy of motion with the potential energy of height. An electromagnetic oscillator (the word harmonic is often lost in the shorthand) exchanges magnetic and electric energies. The basics of an electromagnetic oscillator are shown in Figure 4.31. Charges are initially placed on the two flat surfaces of a capacitor. This produces an electric field and a voltage that accelerates current around the loop. The current does not increase instantly because the increasing current in the loop is opposed by Faraday's law. Eventually, the excess charge is used as it produces more and more current in the loop. When this happens, the electric field energy has been changed to magnetic field energy. After the charges on the capacitor are used, the current does not stop instantly. Faraday's law now slows the current decrease. As the current continues, charges of the opposite sign

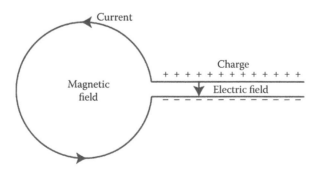

Figure 4.31 A circuit for an electromagnetic oscillator.

are built up on the capacitor. The reversed field associated with this new charge eventually stops the current and reverses its direction. Just as a pendulum swings past its lowest-energy orientation, the electromagnetic oscillator does not stop when the stored charge has vanished. Ideally, the oscillations should persist forever. The natural frequency of oscillation is determined by the properties of the circuit, not the amount of charge placed on the capacitor.

In practice, even old-fashioned electromagnetic oscillators are more complicated than is shown in Figure 4.31. Typically, a larger magnetic field is produced by many loops of wire, and the charge is stored on large surfaces that are folded together to save space. Sometimes, magnetic materials such as iron are part of the oscillator circuit.

4.5.2.6 Resonance and Radio A large resonant response occurs when the frequency of an applied force matches the natural frequency of a harmonic oscillator. Mechanical resonance and examples like singing in the shower were described in Section 3.3.5. A radio contains an electromagnetic oscillator that is resonantly excited by incident electromagnetic waves that match the oscillator's natural frequency. In essence, the radio "hears" only incoming waves at the selected resonant frequency. A radio is tuned to a broadcast signal by adjusting the physical properties of its oscillator so it matches broadcast frequency. In old-fashioned radios, this was sometimes done by varying the area of the capacitor surfaces that carry the positive and negative charges shown in Figure 4.31.

4.5.2.7 Tesla and His Coil Nikola Tesla was a pioneer of electromagnetism. He developed AC motors, the commercial distribution of AC, and much more. The Tesla coil was one of his "flashy" inventions. It produces spectacular lightning-like sparks. The voltages associated with these sparks are difficult to quantify, but they are certainly enormous. Tesla combined the physics of transformers and resonant circuits with enlightened engineering to produce the invention shown in Figure 4.32.

Tesla's advocacy of AC home electrification led to a conflict with Thomas Edison. AC has a clear advantage for long-distance transmission because transformers can reduce current in the transmission lines while delivering the same amount of power. (Power is voltage times current.) Thomas Edison opposed AC, probably because of financial

Figure 4.32 Nikola Tesla and his coil.

reasons. To demonstrate its supposed danger, Edison used AC to publicly execute a number of animals, mostly dogs and cats, but most notoriously an unruly elephant named Topsy.

A strange cult is associated with Tesla. It is difficult to reconcile his brilliant inventions with his fantastic claims (earthquake machine?). Tesla fans have strong opinions on his life, his personality, and his contributions to science.

5

WAVES

5.1 Introduction

Waves are everywhere. Sound waves, radio waves, and light waves are as much a part of the world we know as solids, liquids, and gasses. After an overview of general wave properties with some examples, the descriptions emphasize sound and light waves because without these waves we would be lost. Two of our five senses, vision and hearing, interpret light and sound waves and tell us almost everything we know.

5.2 Common Features of Waves

The geometries of sound and light waves are suggested by the water waves (gravity waves) that can be seen on the surface of a lake or a bathtub. But, the analogy should be approached with caution. In many ways, sound and light are simpler than both water waves and the wave function of quantum mechanics (Chapter 6).

The water waves in Figure 5.1 spread out in circles. A long way from the center, the wave peaks and valleys are nearly straight lines. The waves become nearly "plane waves" with a shape that varies only in the direction pointed away from the wave source. Far from the source, light and sound waves also approach plane wave shape.

5.2.1 Wavelength, Frequency, Speed, Amplitude, and Energy

The simplest wave geometry is the plane wave. The simplest plane wave shape is the "sine wave" shown in Figure 5.2. Any wave shape can be constructed by adding together various sine waves, so the sine wave building blocks of all waves deserve special attention.

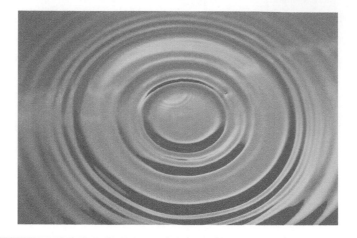

Figure 5.1 Water waves spreading from a central point.

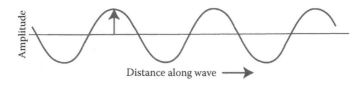

Figure 5.2 The sine wave shape. For water waves, the amplitude is the height. For sound, it is atomic displacement or pressure. For electromagnetic waves, it is the electric field.

Sine waves are characterized by three quantities: wavelength, frequency, and amplitude. The wavelength is the distance between wave peaks. The frequency is the number of times a wave oscillates up and down each second. The amplitude is the height of the wave.

The speed of a wave is the distance one of the wave peaks moves in 1 second. It is related to wavelength and frequency by an important equation.

$$(Wave\ speed) = (Wavelength) \times (Frequency)$$

For example, if the frequency is 6 hertz (cycles per second), then six waves pass a fixed point each second. If the wavelength is 8 meters, then the six wave peaks will have moved 6 × 8 meters in 1 second, and the speed is 48 meters/second.

The speed of a light wave (in vacuum) is a constant regardless of its frequency or its amplitude. Audible sound wave speed is almost constant. The constant speed means plane wave shapes do not change in time. Water wave speeds do depend on the wavelength, which means the wave shapes evolve. This is one reason water waves are more complicated.

Waves contain energy and transmit energy at the wave speed (assuming a constant speed). The energy in each cubic meter of wave is proportional to the square of its amplitude. Multiplying the energy in each cubic meter times the wave speed gives the "energy flux," or the number of joules hitting 1 square meter of surface each second. The most important energy flux is from sunlight. The sun's 1370 watts/square meter keep us from freezing in the dark.

5.2.2 Sound in Solids

The basic mechanism of sound propagation in solids is shown in Figure 5.3. This is a greatly magnified segment of a crystal lattice showing only three of the atoms (spheres) and the atomic forces (springs) that make up the periodic crystal structure.

If atom 1 in Figure 5.3 is displaced to the right, then spring 1–2 is compressed. The compressed spring exerts a force on atom 2. The resulting acceleration and motion of atom 2 compresses spring 2–3, and that carries the motion on to the next atom. For this idealized model, the chain reaction proceeds through the entire lattice of many atoms. The sequence of atomic displacements is not instantaneous because there is a delay between an applied force and the resulting motion. The delays determine the speed of the sound. Light atoms accelerate more rapidly, and strong springs exert larger forces. Thus, it is no surprise that the speed of sound in diamond (about 12,000 meters/second) is 10 times sound's speed in heavy and soft lead (about 1200 meters/second).

5.2.2.1 Polarization The one-dimensional example of Figure 5.3 shows a compression wave by which the atoms move along the direction of the wave velocity. These are longitudinal waves. Transverse

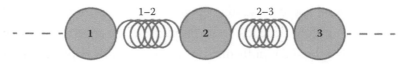

Figure 5.3 Three atoms (spheres) and their interatomic forces (springs). In a solid, many more atoms are part of this chain.

waves cause the atoms to move perpendicular to the wave propagation direction. Transverse waves usually have lower speed because the spring strength for a shear is smaller than for compression. Transverse sound waves cannot propagate through liquids and gases because a lack of rigidity means these materials do not spring back when a shear force is applied. Because transverse seismic waves produced by earthquakes do not propagate through the center of Earth, we know that at least a layer of Earth's core is liquid.

5.2.3 Sound in Gas

Sound propagation in a gas (including air) is similar to the longitudinal waves in solids. In a gas, the motion of the atoms shown in Figure 5.3 is replaced by the motion of thin layers of gas containing many atoms. The springs in Figure 5.3 become regions of compressed gas with larger pressure. A higher-pressure layer accelerates an adjacent layer of gas; this leads to the chain reaction and sound propagation analogous to the sound propagation in solids.

An insightful guess says the speed of sound in gas is roughly the same as the average speed of the molecules (or atoms) that compose the gas. It seems unlikely that sound could move faster than the gas molecules, and there is no obvious reason for sound to lag far beyond the molecular speed. Careful calculations show that the speed of sound in a gas is between two-thirds and three-quarters of the typical speed of one of the gas molecules. Molecular speed increases with temperature because the average molecular kinetic energy is proportional to the kelvin temperature. Thus, the speed of sound in air increases by more than 6% when the temperature increases from a cold day (freezing is 273 K) to a hot day (body temperature is 310 K). Sound speeds in solids or liquids do not show this simple temperature dependence.

Just as sound speed in solids depends on the nature of the solid, sound speed in a gas depends on its properties. The speed of sound in helium, with light and rapidly moving atoms, is nearly triple the speed of sound in air. This high speed explains why a voice is altered to a higher pitch after breathing helium.

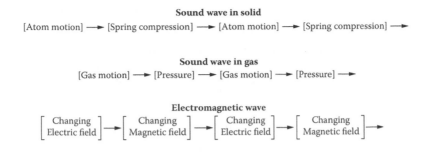

Figure 5.4 Arguments for wave propagation proceeding from the intuitive example of sound in solids to the abstract case of electromagnetic waves.

5.2.4 Electromagnetic Waves

Maxwell's equations have extraordinary power. In particular, they prove that electromagnetic waves (including light) exist, and that they propagate through a vacuum at a constant speed (the fundamental speed of light c). For electromagnetic waves, electric and magnetic fields play the roles of the springs and atomic displacements of sound waves in solids, but the roles of these fields are more symmetric.

The analogy necessary to proceed from sound in solids to sound in gas and on to electromagnetic waves is summarized in Figure 5.4.

The electromagnetic waves are special. Sound waves depend on physical displacement of the medium, either atoms or small regions of gas. Maxwell's equations require no background material. Einstein's special relativity tells us that electromagnetic waves always propagate at the speed of light, even for moving observers.

5.3 Sound Waves in Air

Everything we hear is sound. All we say is sound. For most of us, the pleasure of music and the utility of verbal communication are important aspects of life. Physics explains what we hear and how we hear it.

5.3.1 Sound Wave Physics Background

5.3.1.1 Sound Production Sound is produced by the vibrations of some object that periodically pushes on the air. The frequency of the vibration determines the frequency and wavelength of the sound. The vibrating object can be a drum head, a vocal cord, a loudspeaker, or a volcano. Anything that moves back and forth can produce sound waves. A familiar example is the string on a guitar, violin, or piano. The lowest-frequency vibration of a string, called the fundamental mode, is produced by the motion outlined by the top pair of curves in Figure 5.5. The string oscillates between the solid and dotted curves. The second row of curves shows another way the string can oscillate. For this second case, the frequency is doubled. The same doubled frequency would be obtained if the length of the string was halved, as shown to the left. The bottom row of Figure 5.5 shows the geometries that produce triple the fundamental frequency.

Sound waves with a single frequency are sine wave oscillations of the air pressure, as is shown in the top curve of Figure 5.6. However, a sound wave is often not a sine wave. Most sounds show the periodicity of the fundamental mode, but the time dependence of the oscillation could resemble the sketch shown as the lower curve in Figure 5.6.

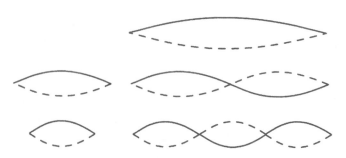

Figure 5.5   Shapes of a vibrating string producing sound with a fundamental frequency two and three times that frequency.

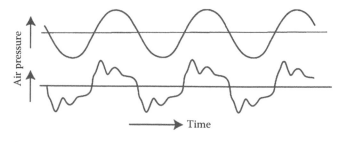

Figure 5.6   The time dependence of two sound waves. The top curve is a sine wave, and the lower curve is a sum of sine waves whose frequencies are multiples of the fundamental frequency.

The complicated wave is a sum of sine waves with the fundamental frequency and multiples of the fundamental frequency.

5.3.1.2 Intensity and the Inverse Square Law   Sound waves carry energy that is proportional to the squared amplitude. Sound intensity is the "energy flux" of the wave: the amount of energy delivered to a 1-square-meter surface in 1 second. A loud sound has an energy flux of 0.01 watts/square meter. This is thousands of times smaller than the energy flux of a typical light source. A bright light feels warm, but loud music will not cure a chill.

Sound often emanates from a compact source, like a violin or a jackhammer. It spreads out more or less uniformly in all directions. The total energy flux through a sphere surrounding the sound source must equal the power producing the energy. The needed formula is *Surface Area* $= 4\pi r^2$, where r is the sphere radius. Figure 5.7 shows that the surface area is proportional to the square of the distance, so the energy flux decreases with the squared distance.

The inverse square law means that if the distance to a noisy object is tripled, the sound intensity decreases by a factor of $3 \times 3 = 9$.

5.3.1.3 Absorption   Sound waves do not propagate perfectly. They lose energy, and this energy is changed to heat in the air. The simplest models predict an absorption rate that is proportional to the square of the frequency. This means the high-frequency part of a sound is eliminated at large distances. Experiments and detailed calculations

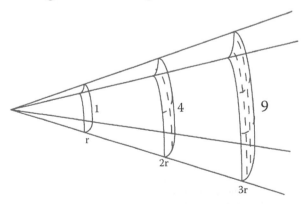

Figure 5.7   The sound intensity decreases with the square of the distance as it spreads out.

show that sound damping is surprisingly complicated. In particular, it varies in strange ways with the relative humidity.

5.3.1.4 Doppler Effect   The frequency of a fire engine siren changes as it passes. It increases on approach and decreases as the fire engine departs. Christian Doppler explained this in an 1842 article on light. His basic idea applies to all types of waves. He said waves are squeezed together when the source is approaching and the waves are stretched when the source is receding, as is shown in Figure 5.8. The squeezed and stretched waves have higher and lower frequencies because frequency and wavelength are inversely proportional. For speeds slow compared to the speed of sound, the formula for the frequency change is

$$\frac{frequency\ change}{frequency} = \frac{approach\ speed}{sound\ speed}$$

This formula is an approximation that can be reliably applied for most everyday cases. However, if speeds are comparable to the speed of sound, the expressions for both moving source and moving listener are changed.

Things become tricky when speeds approach or exceed the speed of sound, as is the case for supersonic aircraft. Figure 5.9 shows that when a source speed surpasses sound speed, no sound can move ahead

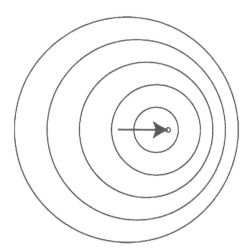

Figure 5.8   The circles represent sound wave peaks produced by a moving sound source. The waves are squeezed in the direction of motion, and shorter wavelength means higher frequency. In the backward direction, the waves are stretched and the frequency is decreased.

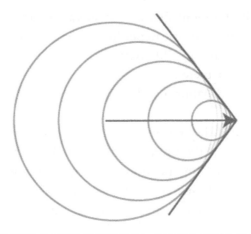

Figure 5.9 The wave peaks are for a sound source moving faster than the speed of sound. No sound can be heard in front of the dark wedge corresponding to the sonic boom.

of the wedge shape moving with the sound source. Sound builds up at the wedge, producing a shock wave or "sonic boom."

5.3.2 Sound We Hear

One can apply the ideas of the physics background to explain our everyday observations of sound effects we hear.

5.3.2.1 Frequency, Pitch, Timbre, and Music We hear sound because the pressure oscillations of the sound waves move our eardrums. Then, a complicated sequence of interactions changes the eardrum motion into our sound awareness.

Pythagoras was the first to scientifically consider musical notes. He observed that simple string-length ratios combined to make pleasing sounds. The octave, resulting from halving the string length and doubling the frequency, as shown in Figure 5.5, is the simplest example. Middle C on a piano is 262 hertz (cycles/second). The speed of sound in air is 343 meters/second. The product of frequency and wavelength is the speed of sound, so the wavelength of middle C is a little more than 1⅓ meters. An octave higher has twice the frequency and half the wavelength.

Healthy young people can hear frequencies from about 20 hertz up to 20,000 hertz, with corresponding wavelengths extending down from 17 meters to 17 millimeters. The 20-hertz low-frequency limit may really be more like a "felt vibration" than a sound, and 20,000 hertz is an optimistic upper limit. The highest frequency most people can sing is around 1000 hertz, corresponding to a wavelength of about a third of a meter.

A sine wave is characterized by its frequency. A typical musical sound is not a sine wave. It will have a complicated but periodic time dependence of the type shown in the lower curve of Figure 5.6. The *pitch* of a note is the inverse of the period, so the two curves of Figure 5.6 have the same pitch even though the lower curve is the sum of sine waves whose frequencies are multiples of the fundamental frequency. Often, "pitch" and "frequency" are not carefully distinguished when a musical note is described.

Most people are pretty good at distinguishing pitch. They can tell which musical note has a higher pitch even when the difference is quite small. A few people have "perfect pitch" (also called absolute pitch). They can identify the pitch of a note with 3% accuracy or better without reference to any other sound. Wolfgang Amadeus Mozart exhibited perfect pitch when he was only 3 years old. "Tone deaf" is roughly the opposite of perfect pitch. The tone deaf have trouble reproducing notes and remembering tunes.

Timbre distinguishes the two curves of Figure 5.6. People can hear timbre differences. The lower curve of Figure 5.6 would have more "character" than the upper curve. The ability to distinguish timbre allows people to identify different musical instruments (e.g., a flute or an oboe). People also have a remarkable ability to distinguish voice sounds. The timbre difference is a major clue that allows one to tell if a song is being sung by John Lennon or Enrico Caruso. Our ability to understand speech depends (in part) on distinguishing the timbres of different vowel sounds, such as *i* and *e*.

5.3.2.2 Sound Intensity and Decibels The sounds we hear every day range from barely audible to uncomfortably loud. A "perceived noise level" is used to classify the intensity of these sound levels on a scale that extends from an almost-undetectable 0 "decibels" (dB) to an uncomfortable 100 dB or even louder.

The physical energy flux in watts per square meter and the perceived decibel scales are very different. Every increase in the decibel level by 10 corresponds to multiplying the energy flux by a factor of

10. Because 100 dB is 0.01 watts/square meter, 90 dB is 0.001 watts/square meter and 80 dB is another factor of 10 smaller, or a measly 0.0001 watts/square meter. Repeating this scaling for a 100-dB range means the ear and its associated complex biology can deal with an enormous range of sound energies. The loudest sound that one should listen to for even a short time (100 dB) carries 10 billion times the energy of the faintest detectable sound (0 dB).

A group of singers illustrates the unusual decibel scale. Increasing the number of singers from four to nine, as shown in Figure 5.10, increases the perceived sound by a modest 3.5 dB. Increasing again by about the same ratio to 20 singers increases the sound only by another

(a)

(b)

Figure 5.10 Nine singers (a) sound about 3.5 dB louder than four singers (b).

Figure 5.11 The jackhammer is a tool with a traditional noisy reputation.

3.5 dB. Some people can notice intensity changes as small as 1 dB. That corresponds to adding one singer to the group of four singers.

The jackhammer of Figure 5.11 is loud. Assume this annoying tool generates 5 watts of sound and assume this power spreads out as a uniform sphere of energy flux. Then, at 5 meters from the jackhammer, the energy flux is 5 watts divided by the surface area of a sphere with a 5-meter radius. Doing the division gives 0.016 watts/square meter, or 102 dB. This is still an annoying noise level. It is no surprise that it is unsafe to stand 5 meters away from a jackhammer for more than 15 minutes. One hopes the person operating this machine has effective ear protection.

Whatever the sound level of the jackhammer may be, the inverse square law means that doubling the distance from the sound source reduces the sound intensity by a factor of four, or 6 dB, which makes it less uncomfortable and considerably less dangerous.

5.3.3 Sound Curiosities

5.3.3.1 Acoustics The enormous variety of wavelengths that can be heard (from 17 millimeters to 17 meters) explains why the acoustic design of a concert hall is so difficult. In principle, the theory of sound is capable of precisely predicting the sound heard in an auditorium. However, the geometry is so complicated that approximations are needed. One tractable approximation works well for wavelength that is small compared to objects in a room. Another approximation is adequate for longer wavelengths. There is no simply applied method that can deal with the enormous range of audible frequencies. This is one of many examples for which increasing computer power is chipping away at the tedious complications of applied physics.

5.3.3.2 Extreme Frequencies   Many animals (dogs, cats, mice) can hear "ultrasound." The prize for achieving the highest frequencies (for animals of reasonable size) probably goes to the bats and dolphins of Figure 5.12. They can produce and hear ultrasonic frequencies exceeding 100,000 hertz. They find the short wavelengths of high frequencies useful for "echolocation."

Elephants, whales, and other animals can communicate with "infrasound" frequencies that are lower than a human can hear. There is little damping at low frequencies, so these sounds can travel great distances. For humans, infrasound of sufficiently large amplitude can cause distress even when a person is unaware of the sound. Some with a conflict-oriented mind-set have wondered if low frequency could be used to disorient an enemy.

Sound waves with frequencies much less than 1 hertz and correspondingly enormous wavelengths are produced by earthquakes and other large-scale natural phenomena. These frequencies are much lower than the rumbling frequencies of elephant communications that extend down to around 10 hertz.

Figure 5.12   Dolphins and bats use high-frequency sound for echolocation.

5.3.3.3 Loud Sounds   It is no surprise that airplane engines and blue whales can make loud sounds. It is a surprise that one of the most prominent ocean noises comes from "pistol shrimp," which are only 3 to 5 centimeters long. These small creatures have an oversize claw whose violent closure makes a sharp "crack" so loud that small fish and other prey are stunned or killed by the sound wave.

Very loud noises have been used in crowd control and in attempts to scare away modern-day pirates (Figure 5.13). This technique has also been employed to keep wildlife away from inconvenient places like airport runways.

One occasionally hears advertisements for 500-watt car sound systems. If this much power really represented the sound inside a car, the noise would be intolerable. Eardrums would burst, and windshields might shatter. Commercial sound system ratings are often exaggerations. At best, the 500-watt rating is the electrical power used to generate the sound, and the conversion from electricity to sound must be inefficient.

An extreme example of long-range propagation of a very loud sound occurred on August 27, 1883. The spectacular Krakatau volcanic eruption annihilated most of the mountain that lay between the islands of Java and Sumatra. Sound was heard in Burma and places in Australia that were thousands of kilometers distant. The blast pressure wave could be detected after it had circled Earth four times. Claims that this was the loudest sound ever heard could well be true.

Figure 5.13   A pirate (perhaps) being discouraged by loud noise.

5.3.3.4 Damping The rapid increase of damping with frequency means high frequencies cannot propagate long distances. But, low frequencies can travel amazing distances under the right conditions. The range of the low-frequency Krakatau explosion is an extreme example. One notices high-frequency damping with thunder. Close to a lightning strike, one hears a sharp "crack" composed of high and low frequencies. Thunder arriving from 1 kilometer away has lost much of its high-frequency part, and only a low-pitched "boom" survives. High-frequency damping means low-budget concert-goers sitting in the back seats will hear the tuba loud and clear, but the piccolo will seem muffled. Because damping depends on humidity as well as frequency, an orchestra playing exactly the same notes in exactly the same place can sound different on dry and damp days.

5.3.3.5 Doppler Effect Buys Ballot (famous mostly for his work in meteorology) was the first to do a careful measurement of the Doppler effect. He placed a group of trumpeters on an open train car that sped along at 18 meters/second (40 miles/hour). This was considered pretty fast for the year 1845. The trumpeters played a note (perhaps a standard A at 440 hertz). A second group of musicians observed the change in pitch, as predicted by Doppler.

Because 18 meters/second is about 5¼% of the speed of sound, the frequency of the trumpets on the approaching train should increase by 5¼%. The difference between two adjacent notes (for example, the A at 440 hertz and A-sharp at 466 hertz) is nearly 6%. Thus, the Doppler shift could be easily identified by the trained ears as almost a half tone. (Today, not all orchestras use the 440-hertz standard.)

Faster-than-sound aircraft produce a sonic boom associated with the shock front sketched in Figure 5.5. The intense noise of a sonic boom can be annoying. The associated pressure can stress an aircraft, and some thought the speed of sound would pose a real limitation on plane speeds. The first person to exceed this "sound barrier" was Chuck Yeager (*The Right Stuff*) in 1947.

Supersonic motion produces some curious effects. For example, assume you could run past a singer at twice the speed of sound. On approach, the apparent frequency of the music would be tripled. After passing the singer, you would hear a backward version of the song, but the notes would be at the correct frequency.

5.3.4 Sound Danger

Old people can no longer hear 20,000-hertz sounds. The deterioration starts young. Few 21-year-olds will notice all the sounds that a 10-year-old can hear. Hearing difficulties become noticeable if the deterioration extends to lower frequencies. The problem is often related to understanding speech. It is the high-frequency part of a pitch that distinguishes the timbre. People with high-frequency hearing loss may have trouble distinguishing a "me" sound from a "my" sound.

Noise is bad for hearing (Figure 5.14). Some surveys indicate that a significant fraction (perhaps one in six) of all US teenagers suffer from some high-frequency hearing loss. It is ironic that young people, who are biologically equipped to hear the faintest sounds, also have an affinity for music that is too loud. Even moderately loud sounds can damage one's hearing over time. To limit this damage, it is recommended that one not be subjected to an energy flux of 0.01 watts/square meter for more than 15 minutes. This corresponds to a sound level of 100 dB. Unfortunately, people in the front row at many music concerts are enjoying more than 100 dB. Extended use of earphones and earbuds is of even more concern. Small battery-powered devices can deliver dangerous sound levels directly to the ears.

Attempts to keep young people from listening to loud noises are not succeeding. A mediocre jingle such as "It ain't no fun when you're 21, but your ears are 81" is not convincing the target audience.

Figure 5.14   No noise sign.

5.4 Light Waves and Geometric Optics

Light is a wave but we cannot see its wave properties because we see with light. We cannot see atoms because they are hundreds of times smaller than light's wavelength. We can see a human hair because its thickness is hundreds of times greater than the wavelength of light. Looking at objects hair size and larger is the realm of geometric optics. Light's wave properties become more prominent as object sizes decrease to the wavelength scale.

Light and vision are intertwined. Euclid and Ptolemy thought light was emitted by the eye. Aristotle thought the essence of physical forms entered the eye. Alhazen, a multitalented Arab or Persian (circa 1000), was probably the first to offer convincing evidence that we see things because light bounces off the objects before entering our eyes. Alhazen performed experiments to justify his ideas. Some call him "the first true scientist."

5.4.1 Speed of Light

The speed of light in a vacuum (and the speed of all electromagnetic waves) is exactly 299,792,458 meters/second because the meter is defined in terms of the distance that light travels in 1 second. Light is so fast that it is hard to notice any delay in its propagation.

Ancient science had no way to measure light's speed, so opinions were really guesses based on philosophy. The following are two famous examples: Aristotle believed light traveled instantaneously; Roger Bacon thought its speed was finite. It is a credit to Galileo that he contemplated a light speed measurement, even though the effort was doomed to failure. He may or may not have performed an experiment he described in 1638. The idea was to station two people, "Anna" and "Bruno," on mountain peaks. Anna flashes her lantern at Bruno, and as soon as he sees it, Bruno flashes back. The time it would take light to travel from Anna to Bruno and back is the distance divided by the speed of light. Light takes only 10 microseconds for the 3-kilometer round-trip. Because human response times are at least 100,000 times greater than 10 microseconds, this experiment only measured Bruno's reflexes. Anna and Bruno could only conclude that light is really fast. They would have no idea how fast.

Ole Romer was the first to obtain a reasonably accurate estimate of light speed in about 1676. Romer worked as an astronomer, so he knew that Jupiter's closest moon, Io, became dark every time it moved behind this giant planet, which is once every 42½ hours. Romer noted that this 42½-hour period was slightly shorter when Earth was moving toward Jupiter than when it was moving away from Jupiter. This change in period is analogous to the frequency shift described by the Doppler effect. Doppler derived a formula that told how frequencies (inverse periods) increase when objects are approaching and decrease when they are receding. Of course, Romer did not phrase his ideas this way because he died nearly 100 years before Doppler was born. Romer's estimate of the speed of light was only about three-quarters of the correct answer because Earth's speed was unclear at that time. Only a century later (after the transit of Venus experiments) were Earth's distance to the sun and its speed determined with reasonable accuracy.

5.4.2 Pierre de Fermat and the Geometry of Light

Mirrors and lenses allow us to manipulate light. Microscopes, telescopes, and eyeglasses help us to better see the very small, the very distant, and everything in between. All these devices apply the principles of geometric optics that accurately describe light as long as physical dimensions are much larger than the wavelength.

It is almost spooky that (nearly) all of geometric optics can be summarized by a single short statement known as *Fermat's principle*:

Light takes a path of least time.

The following examples illustrate the utility of *Fermat's principle* of 1620:

Example 5.1: Straight Lines in a Vacuum

Light's constant speed means the path of least time is also the shortest path, and the shortest distance between two points is a straight line. The path of light is essentially our definition of *straight*. When a carpenter sights along a board and notices that it is warped, Fermat's principle assures him that the board, and not the path of light, is curved. In general relativity, where space is curved, light's path still has a special characterization.

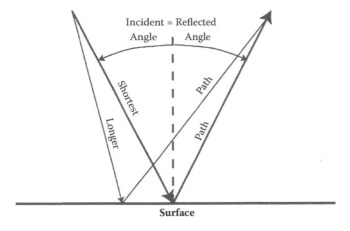

Figure 5.15 Fermat's principle explains why incident and reflected angles are equal.

Example 5.2: Reflection

Fermat was not the first to note that the equality of reflected and incident angles follows from the shortest time principle, as illustrated in Figure 5.15. Hero of Alexandria (inventor of the first vending machine and much more) arrived at the same result 1500 years earlier.

Example 5.3: Refraction

The speed of light is only absolute in a vacuum. Light is slowed when it travels through transparent materials. Light speed in water is three-quarters the speed in vacuum. Light speed in glass is typically two-thirds the vacuum speed. Air also slows light, but only by about 3 parts in 10,000.

Fermat's principle guides light as it travels from one transparent material to another. Minimizing total time means bending the path to limit the distance traveled through the slow-speed region. A quantitative expression for the bending angles is called Snell's law.

Example 5.4: Lifeguard Analogy

A lifeguard and a drowning swimmer illustrate time minimization. Fred (with only his hand showing in Figure 5.16) cannot swim, so it is vitally important that lifeguard Mary reach Fred as quickly as possible. Mary can swim three-quarters as fast as she can run, so it makes sense to run along the beach before diving in

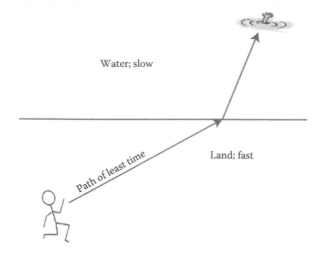

Water; slow

Land; fast

Path of least time

Figure 5.16 Fermat's principle describes the refraction of light as well as the quickest path to rescue a drowning swimmer.

to rescue Fred. Finding the ideal bent path requires Mary to do a time-consuming calculation. Light follows the same path from air to water, and it does so automatically without calculation.

It is a little confusing that minimizing time means paths bend toward the region with slower speed. This is seen in Mary's path; she takes a sharp bend at the water's edge. In a more realistic case where she could run at slower speed through the surf, her path would show a more gradual turn.

Example 5.5: Optical Fiber

Light can travel enormous distances through thin glass fibers like those shown in Figure 5.17 because light bends toward the region

Figure 5.17 A bundle of optical fibers channels light.

of slower speed. An optical fiber slows light most in its center region, so light is confined to the fiber by constantly bending toward the center.

5.4.2.1 Fermat and Sound Fermat's principle can also be applied to sound waves. However, this geometric result is valid only for objects large compared to the wavelength. The ocean is surely large enough. There is a layer of ocean, typically a kilometer deep, where the sound speed is a minimum. Bending toward this slower region means sound can be trapped in the two-dimensional "sofar" (sound fixing and ranging) channel. The channel is most effective for very low frequencies at which the sound wave damping is negligible. The sofar is potentially important for submarine communication. Humpback whales are seen in this region, and they may use its sound-trapping properties to communicate over large distances.

Sound seems to propagate better downwind than upwind. Fermat's principle explains why. Wind speed typically increases with altitude, so when sound travels downwind it will follow a curved path through higher altitudes to save time. Downwind sound bends down to reach the listener, and upwind sound bends up and misses the listener, as is exaggerated in Figure 5.18.

5.4.2.2 Lens Optics The most important lenses are in our eyes. Light entering different points on the lens is directed to a single "focal point" in the back of the eye. Then, biology takes over and allows us to recognize the position of the light source. Fermat's principle explains the lens. Light does not pick one particular least-time path to the focal point because all the paths take the *same* time. Path lengths through the center of the lens are shorter, but time is lost because the central

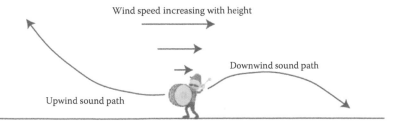

Figure 5.18 Fermat's principle and increasing wind speed with altitude explain why sound propagates better downwind.

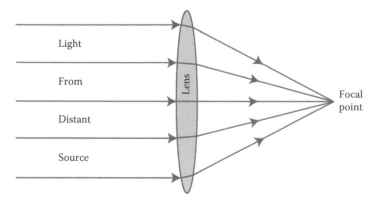

Figure 5.19 A convex lens focuses light from a distant source at a point. All light paths to the focus take the same time, as required by Fermat's principle. If the source is closer, the focal point moves further from the lens. If the source is higher or lower, the focal point moves down or up.

path requires light to move slowly through the thicker part of the convex lens shown in Figure 5.19.

5.4.2.3 Reflecting Telescopes Fermat's principle also explains reflecting telescopes, as shown in Figure 5.20. All light paths from a distant source reflecting off a parabolic mirror converge at a single focal point. As with the lens, these paths to the focal point all take the same time. The concentration of light at the focal point means that even a faint star can be photographed. Another star at a slightly different position

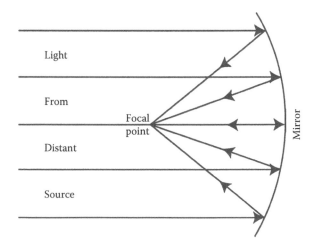

Figure 5.20 A parabolic mirror (nearly spherical) reflects and concentrates light at a single focal point.

will be similarly focused, but its focal point will be slightly displaced so the two stars can be distinguished.

5.4.2.4 Prisms and Rainbows The refraction (bending) of light as it moves from one medium to another has a colorful complication. In a vacuum, all light colors have the same speed. However, blue is typically slowed more than red when it passes through glass or water. Thus, each light color follows a different time-minimizing path, and a glass prism will separate white light (a mixture of all colors) into its colored components. The clarity and magnitude of the separation are exaggerated in Figure 5.21.

The different bending angles for different colors means lenses are imperfect. This problem, called *chromatic aberration*, is present in our eyes, but the effect is small. The absence of chromatic aberration in reflecting telescopes gives them a big advantage over telescopes that use lenses.

Rainbows beautifully illustrate chromatic aberration. A rainbow is produced when sunlight is refracted twice and reflected once by a water droplet. Because the bending is different for different colors, the most likely path, illustrated in Figure 5.22, differs slightly for different colors. Because blue light is reflected further back, the blue band is on the inside of the rainbow.

The light path of Figure 5.22 plays a dominant role in producing rainbow colors. However, there are many other light paths that contribute to the whole picture. The other paths produce additional visible effects, including a second rainbow (shown in Figure 5.23) resulting from a double reflection from the back of the water drop.

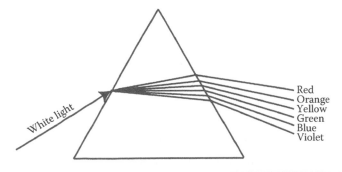

Figure 5.21 Refraction of light by a prism separates colors because light speed in glass depends on the color (frequency).

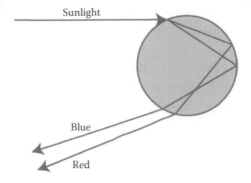

Figure 5.22 Water drops reflect different colored light differently. The different scattering angles can be seen in rainbows when the sun reflects from many drops.

Figure 5.23 The basic rainbow with red on the outside is only part of the geometric optics of light reflecting from water droplets.

5.4.3 Is Light Really a Wave?

Advocates have argued about the particle-wave question for a long time. Isaac Newton said light was a stream of particles. About 100 years later, Thomas Young said light was a wave, and Young's experiment was a milestone in this argument. Today's quantum mechanics blurs the distinction between particle and wave.

Wave effects become important for sizes comparable to the wavelength. For visible light, this distance is much smaller than the breadth of a human hair, but much larger than atomic scales.

5.4.3.1 Diffraction   On very short distance scales, light can bend around corners. For example, after a light beam passes through a very small hole, its wave nature allows it to spread out. It does not continue to travel only in its initial direction.

When light passes through two tiny holes, the light spreads out from each hole. The beams overlap, and the amplitude of the resulting wave is the sum of the two amplitudes. At some points, the peaks in the two waves will coincide, giving large amplitude. At other places, the peak in one wave corresponds to the valley in the other wave. The opposite amplitudes cancel, resulting in no light. Young observed this diffraction using a slightly different geometry.

Young's sketch of a two-slit diffraction pattern is shown as Figure 5.24. The dark lines correspond to wave peaks, and the spaces between the dark lines correspond to wave troughs. One can see lines where the peaks from the two source spots add together and other lines where the peaks cancel the troughs. Making a quantitative analysis of the positions of the peaks allowed Young to obtain a reasonable result for the range of light wavelengths: 400 nanometers for blue and nearly twice that wavelength for red light.

Thomas Young was multitalented. In addition to his contributions to many areas of physics, he was an expert in medicine, language, and music. He played an important role in the translation of the Rosetta stone that led to an understanding of Egyptian hieroglyphics. Young expressed his ideas on the wave nature of light elegantly and forcefully. But, Newton had said light was made of particles, so when

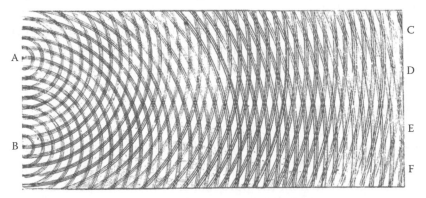

Figure 5.24   Waves from two sources (A and B) add at some places and cancel out at other places. This is Young's original figure that argued for the wave nature of light.

Figure 5.25 Colors reflected from a compact disc are seen when the waves from individual ridges add together to produce maximum amplitude.

Young presented his ideas in 1799, they were not warmly received. It takes time to overcome conformity to a great authority.

Diffraction has many generalizations and important applications. A readily observed example is the colors reflected from the surface of a compact disc (Figure 5.25). A particular color (wavelength) is seen when the amplitudes from the disk's ridges add to maximize the total amplitude. The disk mimics Young's experiment except the light is reflected from many ridges instead of being transmitted through two holes.

A particularly important application of diffraction is seen when x-rays scatter from a periodic array of atoms in a solid. Because x-ray wavelengths are comparable to atomic separations, observing the reflection angle for x-rays of a known wavelength allows one to determine the atomic structure.

5.4.3.2 Waves and Fermat's Principle Diffraction does not invalidate geometric optics. With some effort, one can derive a generalized Fermat's principle from the wave model of light. The generalization gives precise predictions regarding when wave-like effects significantly change the predictions of geometric optics.

5.5 Electromagnetic Waves

Without electromagnetic waves, we could not see because light is an electromagnetic wave. But, visible light is only a small part of the electromagnetic spectrum. Without electromagnetic waves, radios would be silenced. There would be no dental x-rays. We would be lost because our GPS (global positioning system) navigation would fail. Worst of all, cell phones would be dead. Omnipresent electromagnetic waves traveling through empty space at the universal speed limit of 299,792,458 meters/second are as fundamental and as abundant as particles.

5.5.1 Electromagnetic Wave Properties

Electromagnetic waves are special because they can travel through vacuum. Sound waves propagate by moving material (e.g., air molecules), but electromagnetic waves are a self-sustaining combination of electric and magnetic fields that need nothing else. The electric and magnetic fields are perpendicular to the direction of the wave motion. The wave in Figure 5.26 shows a horizontal electric field and a vertical magnetic field. There is another wave that is rotated so the magnetic field is horizontal. *Horizontal polarization* means the electric field is parallel to Earth's surface, and the other orientation is called *vertical polarization*.

Electromagnetic waves interact with essentially everything because electric E and magnetic B fields exert forces on charged particles, and all matter is composed of positive nuclei and negative electrons. The electric force on a charge q is $F = qE$, and the magnetic force on a charge moving with speed v perpendicular to the magnetic field is $F = qvB$. The electric force is always larger than the magnetic force because $B = E/c$ in electromagnetic waves. That means the ratio (magnetic force)/(electric force) is the particle's speed v divided by the speed of light c, and nothing with a mass travels as fast as light.

5.5.2 Example Electromagnetic Waves: Frequencies
Above and Below the Visible

There are neither upper nor lower limits on the frequencies and wavelengths of electromagnetic waves. Pick any wavelength (or frequency),

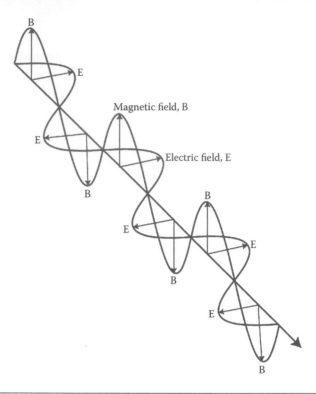

Figure 5.26 The electric and magnetic fields of a horizontally polarized electromagnetic wave.

no matter how ridiculously large or small: There is an electromagnetic wave that fits the bill.

An FM (frequency modulation) radio station broadcast frequency is around 100 million hertz. Because the speed of light is 300 million meters/second, the wavelength of FM signals is about 3 meters. AM (amplitude modulation) radio frequencies are about 100 times lower, so a typical wavelength is 300 meters. If one drives on a bridge with metal superstructures, one notices that the AM radio signal is eliminated but the FM persists. Electromagnetic waves can squeeze past conductors only when the wavelength is smaller than the spaces between the conductors.

The electromagnetic waves of cell phone communication have shorter wavelengths than radio waves. The frequencies extend beyond 2500 megahertz. This is similar to the frequency of a microwave oven. The corresponding wavelength is about 12 centimeters. GPS (global positioning system) wavelengths are a little longer.

Light waves are normally characterized only by their wavelength because the frequencies are so high they do not convey much insight. For example, the wavelength of green light is about 520 nanometers. (One nanometer is 1 billionth of a meter.) It adds little to one's intuition to know that the frequency is 5.77×10^{14} hertz. Green light's wavelength is hundreds of times larger than atomic sizes. That means we cannot see atoms with an optical microscope.

Ultraviolet light has shorter wavelengths and higher frequencies than visible light. It causes sunburn and sometimes skin cancer. Ultraviolet is nominally defined to encompass the wavelength interval from 400 down to 100 nanometers. The distinctions A, B, and C for ultraviolet light correspond to increasingly high frequencies and shorter wavelengths (A ~ 400–315, B ~ 315–280, and C ~ 280–100 nanometers). The shorter waves are potentially more dangerous.

Wavelengths can be much shorter than ultraviolet. Although they are still part of the electromagnetic spectrum, the very short waves have different physical effects and uses. For example, a typical dental x-ray has a wavelength 10,000 times shorter than light (3×10^{-11} meters). This wavelength is comparable to the radius of a hydrogen atom. Under the right conditions, x-rays yield information about atomic structures. Despite lifesaving applications, exposure to x-rays (especially the high-dosage computed tomographic [CT] scans) carries some risk, so x-rays should be used judiciously.

Gamma rays include electromagnetic waves up to the highest frequencies. They are typically produced by radioactive nuclei and astronomical phenomena. They are also dangerous.

5.5.3 Energy and Momentum of Electromagnetic Waves: Heat and a Push from the Sun

The energy sent to us by the sun can be painfully obvious to anyone getting sunburned. In transit from the sun, the radiant energy is stored in the electric and magnetic fields. This energy is proportional to the square of the field strength, and it is carried at the speed of light. A power of 1360 watts for each square meter is directed at us from the sun. Not all of the sun's energy is delivered to us as visible light. Some is infrared (lower frequency) and some is ultraviolet (higher frequency).

If all of the sun's incident power remained on Earth, the heating would soon melt everything. Some of the sunlight is reflected, but about half reaches the surface. Earth does not heat up because it is also radiating, mostly in the form of the invisible infrared electromagnetic waves. A nearly exact equality of incoming and outgoing power keeps Earth's temperature nearly constant. Observations of a warming Earth indicate only a tiny imbalance in this energy exchange.

Electromagnetic waves carry momentum as well as energy. This momentum is associated with a force in the direction of the wave motion. The following suspiciously simplified argument shows that the radiation force is quite small.

When a positive charge q moving with speed v is pushed by a field for a time t, the distance it moves is $d = vt$. The force is qE, and the work done on the charge (force times distance) is $qEvt$. The magnetic field exerts a force qBv on this moving charge, and this force is in the direction the wave is moving. The change in the charge's momentum is the product of the force and time, or $qBvt$. Because $B = E/c$ for electromagnetic waves, the momentum change multiplied by the speed of light is identical to the energy change. This ratio of energy to momentum given to a charge also characterizes the light. The result is

$$(Wave\ momentum) = \frac{(Wave\ energy)}{(Speed\ of\ light)}$$

This means that when light heats an object, it also pushes on it. Although sunlight is trying to push Earth out of its orbit, we need not be concerned. A person with a 1-square-meter cross section standing in the sun feels a force of about 0.00003 newtons. That is much smaller than the force needed to lift a grain of salt.

Light's feeble momentum has observable consequences. Kepler noted that comet tails point away from the sun, as is illustrated in Figure 5.27. The force of the sun's radiation is partially responsible. The sun's radiation also produces a small but detectable correction to the paths of artificial satellites. The radiation force extends throughout the solar system, so some have advocated "space sailing" as a method for traveling to other planets. This is possible if gigantic sails could be erected in space.

Figure 5.27   Comet tails that points away from the sun are aligned by the pressure of the sun's radiation and by a flow of particles called the solar wind.

5.5.4 Polarization of Electromagnetic Waves and Its Applications

The polarization of an electromagnetic wave is determined by the orientation of its electric field. Devices can eliminate one polarization or rotate polarization. Many television and computer screens use these devices to manipulate the polarization and create their displays. People have only the slightest ability to detect polarized light (Haidinger's brush). Bees and octopuses are examples of animals that are sensitive to light's polarization.

5.5.4.1 Polarizer   A polarizer that eliminates the horizontal polarization for wavelengths roughly centimeters long is shown in Figure 5.28. Electromagnetic waves of both polarizations are incident on an array of horizontal conducting wires. The electric field directions for both incident polarizations are shown. The horizontal electric field produces current in the wire. If the separation between the wires is less than a wavelength, the resulting current drains essentially all the energy from the wave, and its transmission is blocked. The vertically polarized wave generates no electric current and passes through the wires with almost no loss.

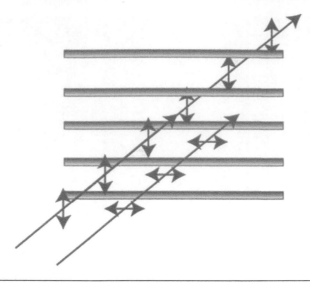

Figure 5.28 The horizontally polarized electromagnetic wave is absorbed by the current produced in the conducting wires. The vertically polarized wave passes through because it produces no current in the wires.

This simple wire arrangement will not work for light because its wavelength is so short. Polarized sunglasses contain a material that responds differently to horizontal and vertical polarizations of visible light. They are microscopic analogues of the wires in Figure 5.28. Light scattered from the sky and water is partly polarized horizontally. Sunglasses that block the horizontal polarization reduce the glare from these reflections.

5.5.4.2 Crossed Polarizers The crossed polarizers shown in Figure 5.29 block all transmission. If a set of vertical wires is placed behind the horizontal wires, waves of both polarizations are absorbed. One can check this with two sets of sunglasses placed in front of each other. Light is blocked when one is horizontal and the other is vertical.

A wire mesh looking somewhat like Figure 5.29 is embedded in the glass panels of microwave ovens. Even though one can see inside, microwave wavelengths on the order of a few centimeters are stopped by the conducting metal grid in the door.

5.5.4.3 Rotation of Polarization Polarization has mysterious properties, as can be seen in simple experiments. One needs just three plastic lenses from polarized sunglasses. When one plastic polarizer is placed

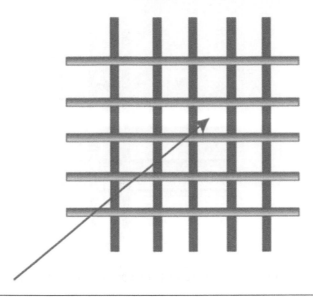

Figure 5.29   Waves are stopped by crossed polarizers because the horizontal wires absorb the horizontal polarization and the vertical wires absorb the vertical polarization.

in front of the other, they can be oriented to block all light. This is the optical analogue of the crossed wires of Figure 5.29. If a third polarizer is placed between the crossed polarizers, it is surprising that its orientation can be adjusted to allow light to pass.

This unexpected three-polarizer transmission is illustrated in Figure 5.30. The third set of conducting wires oriented at 45 degrees is placed between the horizontal and vertical wires. For this case, part of an incident vertically polarized wave is transmitted, but its polarization is rotated by 90 degrees so it exits as a horizontally polarized wave.

The polarization rotation is explained by the decompositions of polarization directions shown in Figure 5.31. The vertically polarized wave is the same as the sum of two waves tilted at a 45-degree angle, and one of the tilted waves can be written as a sum of horizontal and vertical polarizations. For the arrangement of Figure 5.30, the first polarizer allows only the vertical polarization. The second polarizer allows only transmission of the part of the wave tipped clockwise by 45 degrees, and the third polarizer allows only the horizontal part of the tipped polarization.

This decomposition of waves into pieces has analogies in quantum mechanics.

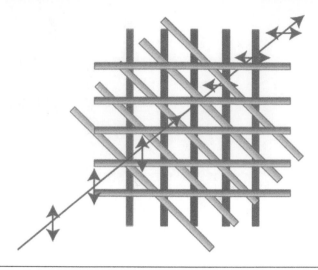

Figure 5.30 Three layers of polarizers can rotate the polarization of electromagnetic waves.

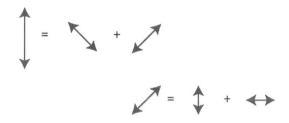

Figure 5.31 Vertical polarization is the same as equal parts of two rotated polarizations. A polarization rotated by 45 degrees is equal parts of horizontal and vertical polarization.

5.5.4.4 Computer Monitors, Televisions, and So On Polarization rotation underlies the physics of liquid crystal display (LCD) screens seen on many watches, calculators, phones, televisions, and computers. Liquid crystals are made up of oriented long molecules, as sketched in Figure 5.32. These molecules are analogous to the conducting wires of Figure 5.28 and will transmit only one polarization. They can also be used to rotate polarization in a manner similar to the wire grids of Figure 5.30. Even though they are oriented, the molecules in these liquids can move about quite freely, and their orientation can be quickly changed by an applied electric field.

When a liquid crystal is placed between two crossed polarizers as is suggested in Figure 5.33, the orientation of the molecules determines whether light is transmitted. The orientation of the liquid crystal is analogous to the orientation of the third polarizer of Figure 5.30.

Figure 5.32 A bit (only seven molecules) of a liquid crystal. The molecules can move as liquids even though they remain oriented.

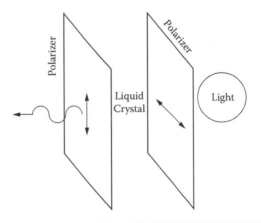

Figure 5.33 Light is transmitted through the crossed polarizers only if the liquid crystal between the polarizers is oriented to rotate the polarization.

Because an electric field can alter the molecular orientations, an applied voltage can turn the light on or off.

The details for real liquid crystal displays differ from the simple example of Figure 5.33. One difference allows liquid crystals to gradually rotate the light polarization so that little light is lost in transmission. For a television screen, there is a sandwich of polarizer and liquid crystal for each pixel on the screen, with the liquid crystal orientation partially controlled by groves in the glass cell containing the liquid crystal. Actually, three cells are needed for each pixel to produce a color image. A lot of engineering is needed to produce these switches on a sufficiently small scale with rapid response times. The light sources behind the pixels are often light-emitting diodes (LEDs).

5.5.5 Electromagnetic Wave Doppler Effect and the Universe

The Doppler shift of light allows us to estimate the age of the universe. It also provides hints about the formation of the universe that are briefly and superficially described in Chapter 8 on relativity.

An approaching red light can appear blue, and a receding blue light can appear red. The Doppler shift is responsible. For either light or sound, an approaching source increases the frequency, but the Doppler shifts for sound and for light have different theoretical underpinnings. Sound is tied to a medium, but light moves through a vacuum. When a light source and an observer are approaching each other, one cannot say which is moving. The relativistic expression for light's Doppler shift cleverly resolves the coordinate system ambiguity. When speeds are small, the Doppler shifts for sound and light are closely analogous.

Doppler understood the importance of his work for light waves even though he did not grapple with relativity and the ambiguous question of who is moving. In 1842, he said:

> It is almost to be accepted with certainty that this [frequency shift] will in the not too distant future offer astronomers a welcome means to determine the movements and distances of such stars.

Doppler was certainly right. The Doppler frequency shift is essential for our understanding of astronomical distances. These distances and velocities are far greater than Doppler imagined, so the relativistic version of the Doppler shift is essential for astrophysics.

5.5.6 Radiation and Scattering of Electromagnetic Waves

5.5.6.1 Radiation Radiation produces electromagnetic waves. It is analogous to the vibrating objects that make sound, but there are obvious differences. Electromagnetic radiation is produced by accelerating charges. Fixed charges produce only static electric fields. Charges moving at a constant velocity lead to both electric and magnetic fields, but no radiation. The power radiated from an accelerated charge is proportional to the square of the acceleration and the square of the charge. The polarization is parallel to the direction of the acceleration.

5.5.6.2 Broadcasting and Receiving Radio stations make their signals by accelerating an electric current up and down an antenna. Because the radiated waves carry energy, external power must be supplied to the antenna. The physics is the same for the production of signals for television and cell phones.

There are many radio stations as well as many other sources of electromagnetic radiation. With all these waves filling space, it is not obvious how a radio or television receiver can pick out the appropriate signal. Government regulations help by requiring all radio stations in a given area to broadcast at different frequencies. Thus, a radio receiver can tune to a particular station using electronics that selects only one frequency for amplification. The trick is resonance. The receiver uses an electrical oscillator (Section 4.5.2) that is an analogue of a pendulum to tune in the desired frequency. When the natural frequency of the oscillator matches the frequency of an incident electromagnetic wave, a very small external signal can excite large-amplitude oscillation in the resonant circuit.

Receiving a signal of the selected frequency is only the first step to obtaining something useful, like music, a TV picture, or someone's message. Information must be embedded in the electromagnetic wave, and this information must be extracted by the receiver. The simplest way to convey information is AM. The amplitude of a radio wave is varied slowly compared to the radio frequency, as is illustrated in Figure 5.34. Electronics inside the radio receiver can detect this signal strength variation and convert it into a sound wave whose frequency reproduces the frequency of the amplitude variation but ignores the high frequency of the incident electromagnetic wave. Modern transmissions use a variety of techniques more complicated than AM to convey information.

5.5.6.3 Scattering and Blue Sky The sky is blue because air molecules scatter the blue frequency of sunlight much more efficiently than they scatter the lower-frequency red radiation. A simple atomic model explains this scattering.

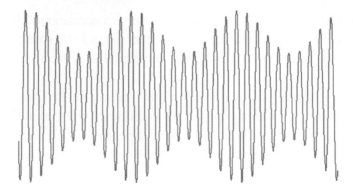

Figure 5.34 An amplitude modulation of a wave.

The electrons in air are tightly bound to their atomic nuclei by the Coulomb attraction. The electric field of sunlight pushes on the electrons. The resulting electron acceleration leads to the radiation of light in a new direction. The energy in this radiation is taken from the incident light, so the diversion of sunlight to radiated light describes the process of light scattering.

The electron acceleration is proportional to the square of the frequency because the quick back-and-forth motion of high frequency is less restricted by the atomic forces than the slower extended motion of low frequency. The blue scattering is further enhanced because the radiation is proportional to the square of the acceleration. The highest-frequency blue we can see is nearly double the lowest visible red frequency. To compare the intensities of blue and red scattering, the two-to-one frequency ratio must be squared twice: once for quadrupled acceleration and once for quadrupled radiation. This means the scattering of deep blue light is nearly 16 times the scattering of the far red. Thus, the sky looks blue. Lord Rayleigh was the first to explain the blue sky. His ideas were refined as knowledge of atomic structure and molecular densities developed.

As shown in Figure 5.35, sunsets are often red because most of the high-frequency radiation has been scattered out of the light beam. By the time it reaches us, the surviving low-frequency red dominates the spectrum.

5.5.7 Vision: The Perception of Electromagnetic Waves

Light and sound are both waves, but we see much differently than we hear. Lenses at the front of our eyes direct light to focal points at the

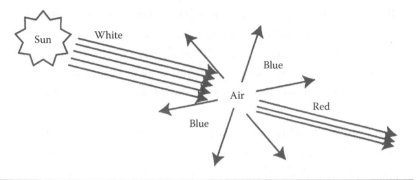

Figure 5.35 Preferential scattering of blue light makes the sky look blue and the sunset red.

back (retina), as is shown in Figure 5.36 The retina is populated with special sensors that tell our brains the position and intensity of the light source. The lens geometry means we can accurately pinpoint a source of light. A comparable precision for sound location is not possible because sound wavelengths are too large to make lenses a practical part of human anatomy. (Middle C wavelength is 1.3 meters.) Carrying about ears that are many meters across so that sound could be focused like light is not a practical option. However, ear shapes do help to roughly detect sound directions. This is especially clear in owls, for which feathers form a roughly parabolic focus toward the ears.

Our ability to distinguish light frequency (or color) is much less precise than our ability to detect different sound frequencies. The highest-frequency blue light we can see is barely double the frequency of the deepest red. By comparison, the range of sound frequencies we can hear is enormous. Young people can hear frequencies 1000 times higher than the lowest-perceptible frequency.

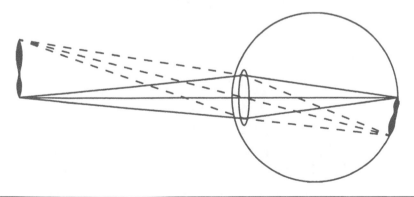

Figure 5.36 Cross section of an eye showing how an image appears at the back of the eye.

The color sensors in our eyes are of three types. Roughly speaking, any color we see is indicated by just three numbers that correspond to the amount of "roughly blue," "roughly green," and "roughly red" light. Because there are more than three frequencies of light, there is plenty of room for ambiguity. Two colors may appear identical even though they are different mixtures of frequencies. This is a real problem for painters and home decorators. On a sunny day, two colors may appear identical, but they can show their hidden differences when the clouds move in. Animals have different color vision from humans. Some see in different frequency ranges, and some have more or fewer than three varieties of color receptors.

Insects have compound eyes like that shown in Figure 5.37. A compound eye may consist of thousands of individual photoreceptor units. The size is 10 micrometers or less. This is only about 10 times the wavelength of visible light. For objects this small, conventional lenses do not work well. Insect eyes are different from ours not because they are more primitive, but because their size requires different physics.

5.5.7.1 Three-Dimensional Images　Our two eyes help us to judge distances. When both eyes aim at an object that is near, we are slightly cross-eyed. Distant objects require parallel eye orientation. Our brain knows how to translate the different eye alignments into a perception of distance. Some movies and televisions use technology that tricks us into believing a flat-screen image is really three dimensional (3-D).

Figure 5.37　An insect's compound eye.

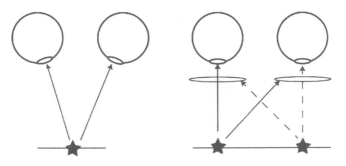

Figure 5.38 Left: An object appears near because our eyes cross to see it. Right: A pair of near objects appears to be a single distant object because each eye sees only one object, and the eyes are not crossed. Special glasses block the light that hides the "reality" of two objects.

The trick shown on the right side of Figure 5.38 allows each eye to see an object at a different position on a flat screen.

For television, a common device flashes the images for the left eye and then the right eye in rapid succession. The viewer wears glasses that allow the image to enter just one eye at a time. A synchronization of the glasses with the images on the screen allows each eye to see its own picture. The two images are arranged so near objects require a cross-eyed gaze. Objects that are supposed to lie in the distance require parallel eye orientation.

For movies, one method of 3-D projection uses the polarization of light. The viewer still must wear glasses; one lens of the glasses transmits only horizontally polarized light, and the other transmits only vertically polarized light. Polarized images for each eye are simultaneously (or intermittently) projected onto the movie screen. As with television, allowing each eye to see something different simulates three dimensions.

The television and movie 3D simulations described here are idealizations and simplifications of evolving technologies. In practice, polarizations more complicated than the basic horizontal and vertical orientations may be used.

5.5.7.2 Hologram   A hologram is an almost magical projection of 3-D images from a flat surface. It shows that the essence of a 3-D structure can be captured in two dimensions. Only bits of hologram physics are described here.

A basic idea of the hologram is a generalization of Young's interference experiment, but with two slits replaced by many holes, as is

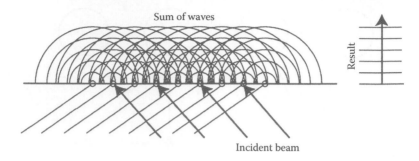

Figure 5.39 The direction of incident light from below the flat surface is indicated by the arrows. The wave crests below the surface are the straight lines perpendicular to the arrows. The little circles are holes where the wave can penetrate. The spherically expanding wave crests propagating from each hole combine to make the upward-directed wave labeled "Result."

illustrated in Figure 5.39. An incident wave is aimed at the surface from below, as denoted by the arrows. This wave feeds through the holes to produce an array of diffracted spherical waves. The web of expanding waves from the holes combines to produce maximum amplitudes on horizontal planes. This generates a wave moving upward, labeled "Result" in Figure 5.39.

The periodic array of holes in the surface of Figure 5.39 has effectively bent the incident wave so that it propagated vertically. This simplest example is obtained because the holes were precisely positioned so wave peaks arrived simultaneously at each hole.

If the holes are placed slightly closer together than is shown in Figure 5.40, one can still obtain a diffracted beam where amplitudes add, but its direction will be bent slightly to the right of vertical. The extra bending is needed to make the path differences from adjacent holes equal to one wavelength. Similarly, when the spacing between holes is increased, the diffracted beam is aimed slightly to the left of vertical.

Variations in the spacing between holes from place to place means the direction of the diffracted beam will change accordingly. An example is shown in Figure 5.40. The wave directions are shown but not the amplitude peaks of the waves. An observer above the surface will perceive the light emanating from a single point below the surface, denoted "Image" in Figure 5.40.

The hologram of Figure 5.40 is primitive. The image appears to be a single point source of light. One can imagine how additional holes

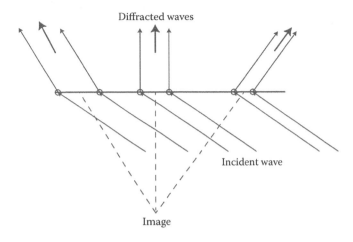

Diffracted waves

Incident wave

Image

Figure 5.40 Narrowly spaced holes produce diffraction bent more to the right, and wider separation gives a diffracted beam aimed more to the left. Holes in the center region correspond to the separation of Figure 5.39, so the diffracted beam is vertical. Observing the light from above gives the impression that it is produced at the point labeled "image."

could be placed in the surface so that light would appear to originate from two different sources. With even more effort, one could generalize to produce an image of a 3-D object.

There remains the intimidating chore of constructing the hologram. Making the little holes by hand is out of the question. Of course, there are tricks to hologram fabrication that are not described here. Suffice it to say that the tricks involve light-sensitive materials, a laser, and half-silvered mirrors. Early hologram production also required a great deal of patience, but constantly evolving technology means applications of hologram-like objects are becoming increasingly common.

6

QUANTUM

6.1 Introduction

The first decades of the twentieth century were the years of the quantum revolution. This revolution differed from many political revolutions. Old ideas were not totally rejected. Newton's laws remain valid for large objects. Maxwell's equations still explain many aspects of light.

The following equation pair shows the close connection between quantum and classical mechanics.

$$F = m \times a$$

$$average(F) = m \times average(a)$$

The first equation is the basis of classical mechanics. It summarizes Newton's laws, where F is the force, m is the mass, and a is the acceleration. The second equation is the quantum-mechanical generalization. It looks almost the same except for the word *average*. Averaging is necessary because particles in quantum mechanics are associated with waves, and one cannot precisely determine the position or motion of a wave. Only the averages are known.

This instructive connection between classical and quantum mechanics is a simplified version of "Ehrenfest's theorem." Ehrenfest was a respected physicist, but his depression eventually led to a tragic suicide.

The equivalence of averaged quantities in quantum mechanics to the corresponding quantities in classical physics tempts one to think quantum physics is not such a big change. This is definitely not the case. Quantum mechanics is conceptually radical. Basic inconsistencies of classical problems are resolved by quantum theory, but they are replaced by the serious conceptual problems of quantum mechanics.

In the following, guideposts to the new ideas of quantum theory are given in terms of six basic rules. They are summarized here and described in more detail in further discussion.

Quantum Rule 1

The energy of any harmonic oscillator is an integer multiple of a basic "quantum energy" ε that is proportional to the product of the oscillator frequency f and Planck's constant h.

$$\varepsilon = hf$$

Quantum Rule 2

A particle is also a wave (called a "wave function") and its momentum p is equal to Planck's constant divided by its wavelength λ.

$$p = \frac{h}{\lambda}$$

Quantum Rule 3

The probability of finding a particle at some position is proportional to the square of the wave function at that position.

Quantum Rule 4

Wave functions and probabilities of the future are determined by wave functions of the present through solutions to the Schrodinger equation.

Quantum Rule 5

Electrons, protons, and neutrons have spin.

Quantum Rule 6

A multielectron wave function must change sign whenever two electrons are interchanged. The same rule applies to protons and neutrons.

These rules are not logically distinct, and they do not form a sensible axiomatic approach to quantum mechanics. They are only an attempt to make complex quantum ideas plausible. One could argue that additional rules are needed. An example is quantum tunneling, which is ignored here, but this concept is described in Chapter 9 on the nucleus.

6.2 What Good Is Quantum Mechanics?

The relation of the "real world" to quantum mechanics is not obvious. Quantum theory says a particle is wave-like, but when one touches a solid object, there is no hint of a wave. Light is supposed to have particle properties, but a flashlight reveals no particles. Heisenberg's uncertainty principle says a simultaneous determination of position and momentum is impossible, but we all know where we are and how fast we are going.

Cars run, electricity works, and the sun rises every morning. All these things appear to happen without any help from quantum mechanics. Knowing about force, heat, pressure, and power is useful in everyday life. Knowledge of wave functions and the uncertainty principle does not seem so important, except for intellectual one-upmanship. One wonders if quantum mechanics only describes things so tiny or so obscure that they can be ignored.

There are, of course, many answers to this negativism. The quantum descriptions that are needed for the microscopic world become noticeable when many small things work together. Much of modern technology uses quantum mechanics. Building a better real world depends on understanding the quantum ideas that underlie everything.

On a more idealistic note, physics seeks to find the ultimate laws of nature. If the search for these laws leads to the obscure, the infinitesimally small, or the ridiculously large, that is the path physics must take.

6.3 Problems with Classical Physics

Cracks in the edifice of classical physics became increasingly troubling in the latter part of the nineteenth century. To illustrate the limitations of the prequantum world, some insoluble problems of classical physics are described here. Solving these mysteries is the happy reward of quantum mechanics.

6.3.1 Atoms

The work of scientists over many years revealed much about the prequantum atom. It was clear that atoms are the building blocks of matter that make up the elements. Scientists of the 1800s discovered more than 50 different atoms, each with its own mysterious properties. In 1869, the Russian Dmitri Mendeleev, whose appearance was as formidable as his brain, introduced the most complete and successful "periodic table." His table noted special relations between different elements and enabled him to correctly predict scandium, gallium, and germanium (elements 21, 31, and 32, respectively, in the table). But, no theory could really explain the structure of the periodic table, whose updated version is Figure 6.1.

An example of a prequantum atomic theory that was doomed to failure was Kelvin's imaginative 1867 knot model. Each knot was supposed to represent a different atom. Examples are shown in Figure 6.2.

Kelvin's knots were not made of strings. They were cores of circling vortices. Thus, the hydrogen atom knot was roughly analogous

PERIODIC TABLE OF THE ELEMENTS

IA																	VIIIA
1 **H** 1.0079	IIA											IIIA	IVA	VA	VIA	VIIA	2 **He** 4.0026
3 **Li** 6.941	4 **Be** 9.0122											5 **B** 10.811	6 **C** 12.011	7 **N** 14.007	8 **O** 15.999	9 **F** 18.998	10 **Ne** 20.180
11 **Na** 22.990	12 **Mg** 24.305	IIIB	IVB	VB	VIB	VIIB	┌──VIIIB──┐			IB	IIB	13 **Al** 26.982	14 **Si** 28.086	15 **P** 30.974	16 **S** 32.065	17 **Cl** 35.453	18 **Ar** 39.948
19 **K** 39.098	20 **Ca** 40.078	21 **Sc** 44.956	22 **Ti** 47.867	23 **V** 50.942	24 **Cr** 51.996	25 **Mn** 54.938	26 **Fe** 55.845	27 **Co** 58.933	28 **Ni** 58.693	29 **Cu** 63.546	30 **Zn** 65.39	31 **Ga** 69.723	32 **Ge** 72.64	33 **As** 74.922	34 **Se** 78.96	35 **Br** 79.904	36 **Kr** 83.80
37 **Rb** 85.468	38 **Sr** 87.62	39 **Y** 88.906	40 **Zr** 91.224	41 **Nb** 92.906	42 **Mo** 95.94	43 **Tc** (98)	44 **Ru** 101.07	45 **Rh** 102.91	46 **Pd** 106.42	47 **Ag** 107.87	48 **Cd** 112.41	49 **In** 114.82	50 **Sn** 118.71	51 **Sb** 121.76	52 **Te** 127.60	53 **I** 126.90	54 **Xe** 131.29
55 **Cs** 132.91	56 **Ba** 137.33	57-71 La-Lu	72 **Hf** 178.49	73 **Ta** 180.95	74 **W** 183.84	75 **Re** 186.21	76 **Os** 190.23	77 **Ir** 192.22	78 **Pt** 195.08	79 **Au** 196.97	80 **Hg** 200.59	81 **Tl** 204.38	82 **Pb** 207.2	83 **Bi** 208.98	84 **Po** (209)	85 **At** (210)	86 **Rn** (222)
87 **Fr** (223)	88 **Ra** (226)	89-103 Ac-Lr	104 **Rf** (261)	105 **Db** (282)	106 **Sg** (266)	107 **Bh** (264)	108 **Hs** (277)	109 **Mt** (268)	110 **Uun** (281)	111 **Uuu** (272)	112 **Uub** (285)		114 **Uuq** (289)				

57 **La** 138.91	58 **Ce** 140.12	59 **Pr** 140.91	60 **Nd** 144.24	61 **Pm** (145)	62 **Sm** 150.36	63 **Eu** 151.96	64 **Gd** 157.25	65 **Tb** 158.93	66 **Dy** 162.50	67 **Ho** 164.93	68 **Er** 167.26	69 **Tm** 168.93	70 **Yb** 173.04	71 **Lu** 174.97
89 **Ac** (227)	90 **Th** 232.04	91 **Pa** 231.04	92 **U** 238.03	93 **Np** (237)	94 **Pu** (244)	95 **Am** (243)	96 **Cm** (247)	97 **Bk** (247)	98 **Cf** (251)	99 **Es** (252)	100 **Fm** (257)	101 **Md** (258)	102 **No** (259)	103 **Lr** (262)

Figure 6.1 The modern periodic table. The numbers in the top of each box are the number of protons in the nucleus.

Hydrogen   Carbon   Oxygen

Figure 6.2 Three examples of Kelvin's model of atoms.

to a smoke ring. The circulating fluid was postulated to be the "ether" that was believed to pervade the universe and support electromagnetic waves. The knot theory never explained anything about atoms, but it did help jump-start mathematical research in knot theory.

Insight into atomic structure was finally supplied by Ernest Rutherford's 1911 model of an atom (based on an experiment by Hans Geiger and Ernest Marsden). This model described the atom as a heavy nucleus surrounded by light electrons. The resemblance of atoms to a miniature solar system, but held together by the Coulomb force instead of gravity, is an appealing image. But, no one could explain why the electrons never fell into the nucleus.

Atomic spectra were another mystery. When atoms are energized, they radiate light at special frequencies. For hydrogen, these frequencies are described by a fairly simple, but unexplained, formula. For all other atoms, the complex spectra were even more mysterious.

6.3.2 X-rays

X-rays are produced when high-energy electrons crash into a target. German physicist Wilhelm Röntgen is usually credited as the discoverer of x-rays in 1895. He was the first to systematically study them, but not the first to notice them. He also gave them the name *x-rays*. It was a good name at the time because *X* usually stands for the unknown. One of Röntgen's early x-ray photos is shown in Figure 6.3.

X-ray images and their possibilities captured public opinion. Stories that women bought lead-lined underwear to protect their modesty from peeping scientists are probably exaggerations because no such undergarments have ever been found.

6.3.3 Radioactivity

Radioactivity was discovered in 1896 by the French scientist Henri Becquerel. While experimenting on a different idea, he discovered that uranium could darken photographic plates. Further work by Becquerel and others (including Marie and Pierre Curie) showed that radioactivity was not simply x-rays. Three types of radiation were identified. They were called alpha, beta, and gamma rays because these were the first three letters of the Greek alphabet and little else

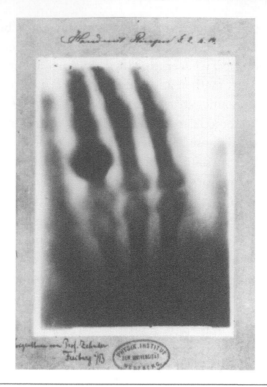

Figure 6.3 This is probably the most famous x-ray image. It was obtained by Röntgen in1895 and shows his wife's hand with a ring.

was known about the radiation. The names persisted even after they were understood.

6.3.4 The Two "Mystery" Gateways to Quantum Mechanics

Neither the heat energy stored in electromagnetic radiation nor the energies of electrons ejected from metals by light beams (the photoelectric effect) could be explained. A proposed resolution of these seemingly unrelated mysteries led to the "photon" concept and the first phase of quantum mechanics.

6.4 Photons

Photons are the solutions to the two mystery gateways to quantum mechanics. Photons were a radical suggestion because they attribute particle properties to light. The observed mysteries and the problems they posed are presented first. The solutions based on the photon model follow. The solutions means we can explain the nature of electromagnetic radiation from heated objects (blackbody radiation), the photoelectric effect and much more.

6.4.1 The Mysteries

6.4.1.1 Mystery 1: The Thermal Energy of Electromagnetic Radiation   Without quantum mechanics, thermal physics runs into perplexing stumbling blocks. Its inability to describe the heat contained in electromagnetic radiation is a striking and historically important example.

6.4.1.1.1 The Observations   Hot objects glow. There is electromagnetic energy in a heated room, and this energy increases rapidly with temperature. The measured electromagnetic energy at each frequency is shown in Figure 6.4. The upper curve, 2T, is at twice the kelvin temperature of the lower curve, T.

The peak in the higher-temperature spectrum appears at twice the frequency of the low-temperature peak. Also, the peak at the doubled temperature is eight times as high as the lower-temperature peak. As a result, 16 times the energy is present at the doubled kelvin temperature.

This electromagnetic energy comes about because hot objects radiate. Things that are not very hot, like people, produce little radiation, and the radiation is mostly restricted to invisible lower-infrared frequencies. Infrared cameras can detect people and other warm objects.

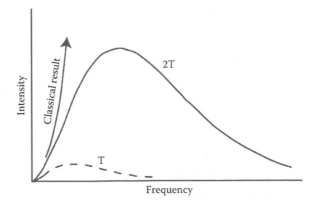

Figure 6.4   The spectrum of electromagnetic radiation (thermal energy at each frequency) at temperatures *T* and *2T* and the classical theory for *2T*.

Figure 6.5 The hot iron radiates intensely, and the frequencies are high enough to be visible as a glow.

These cameras find application in security and military situations. Only when objects are quite hot can we see the radiation. The glow of a blacksmith's work allows the blacksmith to strike when the iron is hot, as is suggested by Figure 6.5.

6.4.1.1.2 The Problem The mystery lies in the inability of classical physics to describe thermal energy and thermal radiation. The incorrect classical result is shown as the curve heading off toward infinity in Figure 6.4.

People knew that all electromagnetic waves are harmonic oscillators, where the oscillators are electric and magnetic fields, rather than a pendulum or the pressure of a sound wave. Regardless of its form, thermal physics assigns an average heat energy kT to every harmonic oscillator, even when the oscillator is an electromagnetic wave. Here, T is the kelvin temperature, and k is Boltzmann's constant, as described in Chapter 3.

This is the core of the problem. There are infinitely many different electromagnetic waves (each with classical energy kT) because there is no shortest wavelength. A wavelength can be a meter, half a meter, a quarter meter, or any other fraction of a meter. This list goes on without end, and each of these wavelengths should contribute kT to the total energy. The absurd theoretical consequence (an infinite energy) and the curve labeled "classical" in Figure 6.4 is called the *ultraviolet catastrophe*.

6.4.1.2 Mystery 2: The Photoelectric Effect

6.4.1.2.1 The Observation When electromagnetic waves hit the surface of a metal such as zinc, electrons are sometimes ejected from the surface of the metal. The electrons can escape only if the frequency of the radiation exceeds a minimum "cutoff" frequency, as is portrayed in Figure 6.6.

6.4.1.2.2 The Problem No one could explain why the electron ejection should depend on frequency rather than the light intensity or the length of time the radiation was applied to the metal. It was also confusing that different frequencies of light were required to eject the electrons from different metals.

The maximum energy of electrons ejected from a metal surface is shown in Figure 6.7. This was also unexplained by any classical theory.

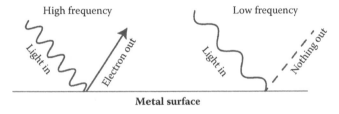

Figure 6.6 Only high-frequency light can eject electrons from the metal surface.

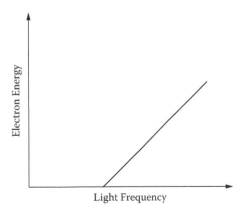

Figure 6.7 The frequency dependence of the maximum energy of an electron ejected from a metal surface by a light beam.

However, data suggested by Figure 6.7 were only obtained by Robert Millikan (the same Millikan who measured the electron charge) after Einstein's explanation. The data of Figure 6.7 are difficult to obtain because even the smallest amount of surface contaminants can change the energies of the ejected electrons.

6.4.2 The Solutions

Max Planck and Albert Einstein led the path into quantum mechanics. Their explanations of electromagnetic radiation (Planck) and the photoelectric effect (Einstein) make sense if one accepts the following rules:

Quantum Rule 1

The energy of any harmonic oscillator is an integer multiple of a basic "quantum energy" ε that is proportional to the product of the oscillator frequency f and Planck's constant h.

$$\varepsilon = hf$$

The fundamental constant h that determines the allowed energies of a harmonic oscillator is always called Planck's constant. This rule, called *energy quantization*, is a foundation of quantum mechanics. Energy quantization is perplexing and nonintuitive. An equation that explains experiments at the cost of common sense is a hallmark of quantum mechanics. There is no obvious reason why an oscillator is allowed to have energies, 0, *hf, 2hf, 3hf*, and so on while it is forbidden to have an intermediate energy, such as $(1/3)hf$. Although Planck never accepted all of quantum mechanics, he apparently realized that his ideas were the beginning of a revolution. His son, Erwin, reported that his father told him in 1900: "Today I have made a discovery which is as important as Newton's discovery." At other times, his comments were much more modest.

6.4.2.1 Solution to Mystery 1: The Thermal Energy of Electromagnetic Radiation Quantum rule 1 can be applied to light because electromagnetic waves are really harmonic oscillators. When one multiplies Planck's constant by the very high frequency of visible light, the energy is comparable to other atomic-scale energies. For example, the quantum energy hf of green light is 2.5 electron volts.

Quantum rule 1 puts a cap on the thermal energy stored in high frequencies of electromagnetic waves. Thermal physics wants to give kT of energy to every oscillator, but quantum rule 1 requires this energy to be a multiple of hf. When the quantum energy hf is greater than the thermal energy kT, there is not enough heat. The most likely outcome is zero thermal energy. As is suggested in Figure 6.8, the high-frequency oscillations are "frozen out" when the thermal energy is significantly smaller than the quantum energy. The elimination of the high-frequency energy saves the world from an ultraviolet catastrophe. Planck determined his constant by matching his theory with the experimental results of Figure 6.4.

The freezing out of high frequencies and the curves in Figure 6.4 are manifest in everyday life and in stars. The quantum energy hf of green light is about 100 times the thermal energy kT of a person whose temperature is 310 K. People do not glow with a green color because the visible photons are frozen out. People only emit low-frequency infrared light. A stove is hot when it starts to glow red. The glow is red because red has a lower frequency than other colors. Higher-temperature produces much more radiation, and the average frequency of the radiation increases with the temperature. When something is really hot, the red glow becomes white. When a blacksmith notes the color of his work and sees that the iron is hot, he is applying principles of quantum physics to judge the temperature. Astronomers apply the blacksmith's technique, but with precision instruments designed to accurately determine the temperature of a star by the color of the light

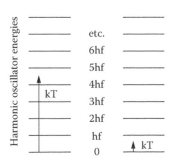

Figure 6.8 On the left, the thermal energy kT is larger than the quantum energy hf. The average thermal energy will be about kT. On the right, the thermal energy is less than the quantum energy, and the oscillator energy will usually be zero.

it emits. A precision instrument is not needed to tell one that blue stars are much hotter than red ones.

6.4.2.2 Solution to Mystery 2: The Photoelectric Effect   Albert Einstein said the quantized energy of light should be taken as a physical reality and not just a trick to explain away the ultraviolet catastrophe. He used Planck's quantization to describe the photoelectric effect by extending quantum rule 1.

Extended Quantum Rule 1

When an electromagnetic wave with frequency f interacts with a metal, it transfers one energy quantum hf to one of the electrons in the metal.

Electrons are normally held inside a metal by a surface energy barrier, denoted ϕ. When light transfers its energy quantum to an electron, that electron can escape if the quantum energy is large enough to surmount the barrier ($hf > \phi$). The barrier energy of many common metals is between 4 and 5 electron volts. The energy quantum for green light is 2.5 electron volts, so green light cannot eject electrons. It takes ultraviolet light with roughly twice the frequency of green light to free the electrons. When hf of the light exceeds the barrier energy, the extra energy can be given to the escaping electron, as is shown in Figure 6.7.

6.4.3 Photon Properties

Quantum rule 1 and its extension are given linguistic legitimacy by the invented word *photon*. Naming something (a photon in this case) is not really an explanation, but it helps to make quantum mechanics a little more palatable. The assertion that light traveling through space is both a wave and a flow of particles means photons have the following unusual properties:

1. The photon's energy is hf.
2. Photons can appear and disappear. They are created by hot surfaces. They are destroyed when they give their energy to an electron.
3. Photons (in vacuum) always move at the speed of light.

4. Photons have no mass. However, they have energy. This is a problem because $E = mc^2$ suggests a mass. This famous formula relating mass and energy and its extension to photons are described in Chapter 8 on relativity.
5. Massless photons also have momentum p because light has momentum. The relation $p = E/c$ for light is also true for photons.
6. Groups of photons can exhibit wave properties such as interference and diffraction.

This curious superposition of light's wave and particle properties is called *wave-particle duality*. Quantum mechanics asks us to believe in this duality because it is the only way to explain experiments.

The photon picture of light allows one to depict the photoelectric effect in terms of a simplified version of a Feynman diagram, as shown in Figure 6.9.

The Feynman diagram suggests how the electron ends up with the additional energy of the photon. To conserve both energy and momentum, the electron must bounce off something else. In empty space, an electron cannot absorb all of a photon's energy. This is fortunate because we need light from our sun, and we appreciate the beauty of stars. Feynman diagrams are more than the suggestive cartoon of Figure 6.9. They are backed up by instructions that allow one to calculate precisely how often this process will take place.

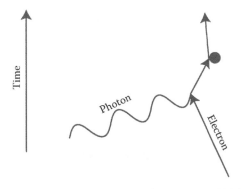

Figure 6.9 A photon gives its energy to an electron and disappears. The electron bumps into something else, denoted by the dot, so both energy and momentum can be conserved in this process.

6.4.4 Other Harmonic Oscillators

Sound waves in solids are harmonic oscillators. Quantum rule 1 applies to these oscillators as well. The quantized sound vibrations of solids are called *phonons*. Quantum rule 1 means lattice vibrations (like light) are frozen out at low temperature. Thus, atoms in a crystal stop vibrating (almost) as the temperature is lowered toward absolute zero.

The vibrations of atoms are also quantized. At a lower temperature at which an atomic vibration frequency multiplied by Planck's constant is much less than kT, the vibration is again frozen out. For many molecules, this means their thermal vibration energies kT appear only at temperatures above room temperature. Before quantum mechanics, the freezing out of harmonic oscillators and the low-temperature thermal energies of solids and molecules had no explanation.

6.5 Particles and Waves

6.5.1 de Broglie Hypothesis

If light waves are particles (photons), perhaps particles are waves. This was the 1924 suggestion of Louis-Victor-Pierre-Raymond, 7th duc de Broglie. More precisely, de Broglie postulated the following:

Quantum Rule 2
A particle is also a wave, and its momentum p is equal to Planck's constant divided by its wavelength λ.

$$p = \frac{h}{\lambda}$$

Planck's constant is tiny. The momentum ($p = mv$) of ordinary moving objects is large. As a result, de Broglie's formula gives ordinary objects unmeasurably small wavelengths. The electron is different. Its very small mass means its momentum is small even when it is moving rapidly. Despite its small mass, the value of Planck's constant still yields a very small electron wavelength. But, the electron moves in a very small space (the size of an atom). Thus, the electron wavelength demanded by the de Broglie hypothesis is important for atomic physics.

6.5.1.1 Logic Behind Quantum Rule 2 The guess made by de Broglie can be partially justified from three equations describing a single photon:

$$\varepsilon = h \cdot f$$

$$p \cdot c = \varepsilon$$

$$\lambda \cdot f = c$$

The first equation is energy quantization (quantum rule 1). The second says the relation between momentum and energy applies to photons as well as electromagnetic waves. The third says the product of frequency and wavelength for a light wave (or a photon) is the speed of light c. These equations and some algebra yield $p = h/\lambda$ for a photon. de Broglie argued that this relation should apply to all particles, but the details of his reasoning used relativistic formulas.

The particle-wave duality is just as confusing for an electron (or anything else) as it is for light. The flippant answer to this confusing duality is that sometimes it acts like a particle and sometimes it acts like a wave. Debates on this conundrum continue.

6.5.2 Quantized Angular Momentum

Planck's constant divided by 2π is the quantum-mechanical unit of angular momentum. This can be illustrated for a particle confined to move on the circle shown in Figure 6.10. Four wavelengths are wrapped around the circle. The circumference of the circle is $2\pi r$, where r is the circle radius. Thus, the wavelength is $\lambda = 2\pi r/4$.

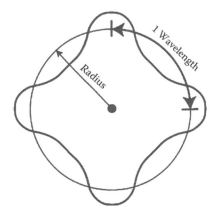

Figure 6.10 A particle moving on a ring. The ring circumference is equal to four wavelengths. This picture is also a simple model of a hydrogen atom with the center dot representing the proton.

Quantum rule 2 says this particle has momentum $p = h/\lambda$. The angular momentum of an object moving in a circle is the product of the momentum p and the radius r. For the particle in Figure 6.1, this means the angular momentum is four times $h/(2\pi)$. There is nothing special about wrapping four wavelengths around the circle. The number of waves could be any integer, but it cannot be a fraction or the wrapping would not work. Thus, one arrives at a fundamental quantum principle called *quantization of angular momentum.*

$$angular\ momentum = integer \times \frac{h}{2\pi}$$

This result is general. Even if the particle is not constrained to move on a ring, rotational symmetry alone means the angular momentum is quantized in multiples of $h/(2\pi)$. Angular momentum quantization is a key to describing the quantum mechanics of atoms.

6.5.3 Hydrogen Atom

6.5.3.1 Hydrogen Atom Energies   Hydrogen is the simplest atom. It is composed of a single proton and a single electron. Its simplicity was a doorway to quantum mechanics. The most accessible property of the hydrogen atom is the series of "quantized energies," one for each value of a "quantum number" n.

$$E_n = -\frac{13.6\ electron\ volts}{n^2}; \ n = 1, 2, 3, \ldots$$

This formula is just a shorthand way of saying that it takes 13.6 electron volts to separate the electron from the proton when hydrogen is in its lowest-energy state ($n = 1$). But, hydrogen also has higher-energy states. When hydrogen is in its first excited state corresponding to $n = 2$, only one-quarter of 13.6 electron volts is needed to free the electron, and for the second excited state, the required separation energy is only a ninth of the 13.6 electron volts. In principle, this sequence of energies goes on forever. The energies are sketched in Figure 6.11. They are negative because work must be done to free the electron from the pull of the proton. Only energies with an integer value of n are allowed. This quantization of the energies means the average distance between the electron and proton is also quantized.

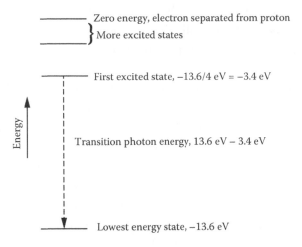

Figure 6.11 The energy levels of the hydrogen atom. A transition from the first excited state to the lowest-energy state produces a photon with energy equal to the energy decrease.

6.5.3.2 Measurement of the Hydrogen Atom Energies Hydrogen's quantum energies were obtained from experiments and the formula for the energy of a photon. A hydrogen atom can "jump" between the different energies shown in Figure 6.11. Total energy is conserved, so a photon is created when the atom loses energy, and a photon is absorbed when the atomic energy increases. For example, if the atom jumps from the $n = 2$ energy level down to the $n = 1$ energy level, the frequency of the emitted photon is determined by energy conservation.

$$hf = \frac{13.6 \ electron \ volts}{1^2} - \frac{13.6 \ electron \ volts}{2^2}$$

Measurement of this (ultraviolet) photon frequency determines the difference between the $n = 2$ and $n = 1$ energies. Many additional photon frequency measurements are needed to reveal the pattern that describes all the hydrogen atom-binding energies.

6.5.3.3 Bohr's Hydrogen Atom Niels Bohr was the first to obtain correct expressions for the hydrogen atom energy levels. His extraordinary physical insight allowed him to succeed a decade before de Broglie's wave-particle duality. Because Bohr did not accept photons, his imaginative reasoning is all the more impressive. Hindsight and help from de Broglie allows the simplified but plausible Bohr-like explanation of the energy levels that follows. Many view these simplifications as dishonest and confusing because electrons really do not follow classical orbits. This dishonest path to a correct result is presented apologetically.

In the spirit of Bohr, imagine an electron orbiting a proton in a classical circular orbit so the hydrogen atom resembles Figure 6.1 with the dot in the center representing the proton. A classical analysis of the orbital motion shows that the binding energy is inversely proportional to the separation between electron and proton. Classical mechanics also shows that this separation is proportional to the square of the angular momentum. Quantum mechanics now enters the picture because the angular momentum is quantized. It must be a multiple of $h/(2\pi)$. The quantized angular momentum thus translates into quantized energy levels. Bohr obtained a formula for the 13.6-electron-volt energy as well as the excited-state energies. He showed that these energies can be expressed as a combination of three fundamental constants: the charge of the electron, the mass of the electron, and Planck's constant (or e, m, and h, respectively).

Bohr also gave a theoretical explanation of atomic size. The Bohr radius $a_0 = 5.29 \times 10^{-11}$ meters is a characteristic atomic radius that can also be expressed in terms of these three fundamental constants. The Bohr radius is very small. One billion Bohr radii placed end to end are only a little more than 5 centimeters long.

6.5.3.4 Beyond the Bohr Model   Despite its success, Bohr's model is not easily extended to more complicated atoms. The Bohr atom tempts one to accept some unjustified simplifications. The electron wave is not really constrained to move on a circle as suggested by Figure 6.10. It is not even constrained to move in a plane. It is better to visualize the electron wave as a cloud surrounding the proton. For the lowest-energy state, the cloud is concentrated near the proton but extends past the Bohr radius. Its spherical shape is sketched in Figure 6.12.

6.5.4 Probability and Uncertainty

6.5.4.1 Wave Function   The cloud that describes a particle is really a wave in three dimensions. It is called a *wave function*. Wave functions

Figure 6.12 The three-dimensional electron cloud surrounding a proton. The shading is proportional to the probability of finding the electron. This probability is largest at the proton's position.

are not required to have the shape of a simple sine wave, as is clear from the wave function (or cloud) sketched in Figure 6.12. Because an electron is described by a wave function, one does not know exactly where it is. One can only tell where the electron will probably be found using quantum rule 3.

Quantum Rule 3
The probability of finding a particle at some position is proportional to the square of the wave function at that position.

6.5.4.2 Heisenberg's Uncertainty Principle The vagueness of quantum mechanics is neatly expressed by the uncertainty principle. The very small size of Planck's constant explains why we do not notice the uncertainty.

An example of a wave function that might describe a particle is shown in Figure 6.13. The most likely place to find the particle is at the center because the square of the wave function is largest at this point. However, the particle also could be detected away from the origin, but with a smaller probability. Thus, there is an uncertainty in the position, and that uncertainty is roughly the width of the curve of Figure 6.13.

Because this wave is not a sine wave, it is difficult to determine a wavelength. The momentum is Planck's constant divided by the

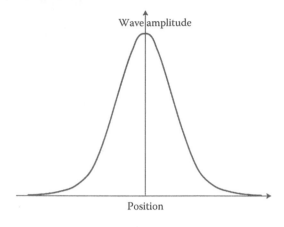

Figure 6.13 An example wave function.

wavelength, so the momentum is also imprecisely known. With only one wave peak, one can only crudely guess that the wavelength uncertainty is about the same as the width of the peak. These uncertainties suggest two approximate relations.

$$(\text{position uncertainty}) \; \textit{is roughly} \; (\text{wave width})$$

$$(\text{momentum uncertainty}) \; \textit{is roughly} \; \frac{h}{(\text{wave width})}$$

Multiplying these two guesses means the product of position uncertainty and momentum uncertainty is roughly equal to Planck's constant. A precise definition of the uncertainties solidifies this result, known as the *Heisenberg uncertainty principle*.

$$(\text{position uncertainty}) \times (\text{momentum uncertainty}) \geq \frac{h}{4\pi}$$

The greater than or equal to symbol ≥ is used because oddly shaped waves can have greater uncertainties.

6.5.4.3 Consequences of the Uncertainty Principle   The uncertainty principle is perplexing. It means that when one tries to determine the position of an object on a very small scale, the act of measurement will give that object a large and unknown momentum. Similarly, a precise determination of the momentum must sacrifice knowledge of the position.

In classical mechanics, the position and momentum of an object must be specified before its future motion can be determined. By denying us simultaneous knowledge of position and momentum, the uncertainty principle says classical motion cannot be determined. This would appear to say the future cannot be predicted. Some have associated the uncertainty principle and its indeterminacy with philosophical concepts like "free will." Most people familiar with quantum mechanics would not agree that the uncertainty principle resolves any religious arguments. However, questions about causality in the context of quantum mechanics are fully as challenging as the classical philosophy.

6.5.4.4 Why Matter Takes Up Space   The uncertainty principle is not a mistake, and it will not go away. That is a good thing because it explains why atoms have size—even when the angular momentum vanishes. Suppose one pushes an electron very near a proton. The proton and electron attract each other, so the potential energy decreases. But, energy comes in two parts. The uncertainty principle says that an attempt to confine the electron in a small space increases its average momentum, and the kinetic energy is proportional to the square of the momentum. When compressed, the increased kinetic energy overwhelms the decrease in potential energy; this keeps the atom from collapsing.

When the hydrogen atom energy is minimized (one says it is in its "ground state"), the electron has zero angular momentum. It does not have a classical orbit. It is a spherically symmetric wave surrounding the proton, as illustrated in Figure 6.12. Atomic dimensions are equal to the Bohr radius because this is the size that minimizes the sum of kinetic and potential energies.

The Bohr radius is not zero. That is important. It means atoms cannot collapse to a point. Furthermore, atoms cannot be at the same place at the same time. The atomic model and electrostatics offer a rudimentary explanation. Solids are hard and impenetrable because the electrons in one atom repel the electrons in a neighboring atom. The positive nuclei also repel. Rigidity results because atoms cannot be squeezed too closely together. When the carpenter in Figure 6.14 hits his thumb with a hammer, he should not blame his hammer for the pain. He should blame quantum mechanics and Coulomb's law.

Figure 6.14 The thumb and hammer cannot be at the same place at the same time because of quantum mechanics and Coulomb's law.

6.5.4.5 Hydrogen Molecule The simplest example that shows how atoms take up space is the hydrogen molecule. The protons and the electrons in the two hydrogen atoms repel each other, but the protons attract the electrons. The balancing of these forces shown in Figure 6.15 means the atoms attract each other at a large distance, but they strongly resist being pushed too close together. The resulting molecule has it lowest energy when the separation between the two protons is about 1.5 times the quantum-mechanical Bohr radius. Trying to squeeze them closer together takes a large force. This result generalizes to solids composed of a large number of more complicated atoms.

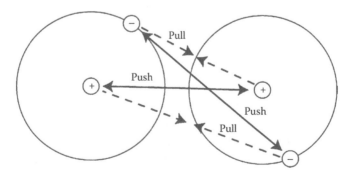

Figure 6.15 The forces that determine the separation between the two hydrogen atoms in a hydrogen molecule. This is a simplification because electrons are really described by waves.

6.5.4.6 Zero-Point Energy Just as the uncertainty principle keeps atoms from collapsing, it forbids harmonic oscillators from becoming perfectly still. Classically, a pendulum (or any harmonic oscillator) has a minimum energy when both its displacement and momentum are exactly zero. But, the uncertainty principle forbids this simultaneous specification of position and momentum, so the oscillator has (on average) a small amount of momentum and displacement. The harmonic oscillator energy can be as small as half the energy quantum (*hf*/2), but it cannot vanish. This remnant, called *zero-point energy*, is a problem. Electromagnetic waves are harmonic oscillators, and the uncertainty principle tells us there must be zero-point energy in every electromagnetic wave. Because there is an infinite number of electromagnetic waves with differing wavelengths, this means there is an infinite amount of energy contained at every point in space. Attempts to sweep this infinity (and related, but more complex, infinities) under the rug have not been totally successful. This is an especially difficult problem when one tries to blend quantum theory with modern ideas about gravity.

6.5.4.7 Schrodinger's Cat It is possible for Schrodinger's cat to be both alive and dead when it is described by a wave function that is a sum of a "live part" and a "dead part." Even though Figure 6.16 seems unconvincing, this weird conclusion is consistent with the formalism of quantum mechanics.

Figure 6.16 Some believe quantum mechanics says a cat can be both alive and dead.

If quantum mechanics says a cat can be both alive and dead but the two cat states cannot know of each other, there is a question of reality. From the live cat's point of view, is the dead cat real?

The logical extension of two cat forms is even more disturbing. People who view this truly schizophrenic cat should also be described by wave functions made of two pieces. One piece would see a live cat, and the other piece would think the cat was dead. The possibility that a single human can be described by a wave function with two very different states of mind is surely counterintuitive.

Philosophical questions related to Schrodinger's cat will not be answered here.

6.5.5 Schrodinger Equation

The Schrodinger equation appeared only about 3 years after the de Broglie hypothesis. This equation is the workhorse of nonrelativistic quantum mechanics. It can be interpreted as an expression for the total energy:

Total energy = Kinetic energy + Potential energy

This energy identity is transformed into an equation for the wave function. The time dependence of the wave is obtained from the relation *Total energy* = hf because the frequency f is associated with variation in time. Mathematics extracts a wavelength λ and momentum $p = h/\lambda$ for waves of any shape, and the kinetic energy is determined by the momentum. The potential energy depends on the problem. For the hydrogen atom, it is the Coulomb energy associated with the force of attraction between the negative electron and the positive proton. The Schrodinger equation can be generalized to any form of potential energy and to wave functions of any shape.

Example accomplishments of the Schrodinger equation are the following:

1. A derivation of all the hydrogen atom energies, corresponding wave functions, and the size of the hydrogen atom.
2. A derivation of the energy levels of the harmonic oscillator, showing that energy changes are multiples of hf as stated in

quantum rule 1. The Schrodinger equation also obtains the troubling zero-point energy $hf/2$.
3. Solutions to the Schrodinger equation mean the causality of classical mechanics is resurrected in a vague wave-like form, as expressed in quantum rule 4.

Quantum Rule 4
Wave functions and probabilities of the future are determined by wave functions of the present through solutions to the Schrodinger equation.

The mathematics needed to solve the Schrodinger equation is tedious. Some have said this equation should not be named after Schrodinger because he did not derive it and he could not solve it. It is true that the historical origins of the Schrodinger equation are a little bit foggy (and a little bit racy), and it was obtained partly from physical insight instead of rigorous mathematics. Also, Schrodinger needed help from mathematician Hermann Weyl to obtain the hydrogen atom energies and wave functions.

The Schrodinger equation describing a single electron is not the end of the story. It can be generalized to many particles. One should not be overly enthusiastic about this transition to many particle quantum mechanics. As described in Section 6.8, writing down a many-body Schrodinger equation and solving it are two entirely different issues.

6.6   What Is *137*?

The hardest and most profound questions are often stated simply. The number 137 appears again and again in quantum mechanics. The question is, Why is this number 137?

6.6.1 Curious Facts

The relativistic energy of the electron mass mc^2 divided by twice the hydrogen atom ground-state energy is

$$(137)^2 = \frac{mc^2}{2 \times 13.6 \; electron \; volts}$$

A closely related ratio is the speed of light c divided by the speed v of the electron (appropriately averaged) in this hydrogen atom:

$$137 = \frac{speed\ of\ light}{averaged\ electron\ speed}$$

The energy of a photon with a wavelength equal to the circumference of a circle of radius r divided by the potential energy of two electrons separated by the same distance r is

$$137 = \frac{photon\ energy}{Coulomb\ energy}$$

Corrections to basic quantum theory (quantum electrodynamics) are commonly expressed as a series of increasingly complicated adjustments. These corrections become smaller and smaller (one hopes) only because each additional term is multiplied by another factor of 1/137.

6.6.2 A Universal Number

In convenient "natural" units,

$$137 = \left(\frac{h}{2\pi}\right)\frac{c}{e^2}$$

As before, h is Planck's constant, c is the speed of light, and $-e$ is the electron charge.

If an arriving alien spaceship was decorated with 137 dots, we would know the inhabitants of the ship were aware of modern physics: 137 is an absolute number, incorruptible and independent of language and the system of units.

One uses 137 as an approximation to 137.03599907. One can imagine the amount of experimental and theoretical work and the collaborations of many scientists needed to obtain this extraordinarily accurate determination.

6.6.3 No Explanation

Many have spent countless hours trying to understand the significance of 137, to no avail. The brilliant and eccentric physicist Wolfgang

Pauli developed a strange relationship with the equally unusual psychotherapist Carl Jung, partly because of Pauli's preoccupation with 137. Strangely, when Pauli was dying, his last hospital room was numbered 137.

Richard Feynman, a famous physicist with different eccentricities, commented on 137.

> All good theoretical physicists put this number up on their wall and worry about it.... It is one of the greatest damn mysteries of physics: a magic number that comes to us with no understanding by man. You might say the "hand of God" wrote that number.

6.6.4 Anthropic Principle

We are lucky. Our universe would not be habitable if 137 were 127 or 147. Does this mean the hand of God wrote that number to allow us to exist? There are other fundamental numbers in addition to 137. These numbers all appear to have taken on special values that make us relatively comfortable. Perhaps there are other universes with different values for 137, and people appear only in universes with "convenient" values of the fundamental constants.

6.6.5 Numerology

Some speculations about 137 are stranger than others. Two examples give the spirit of the quest:

1. It has been claimed that the number of particles in the universe can be obtained by multiplying 2 times itself 137 times and then squaring the result.
2. Kabbalah (associated with esoteric teaching) written in Hebrew letters is

$$\text{קַבָּלָה}$$

   Numbers in Hebrew are traditionally denoted by letters, with each number corresponding to a different letter. The four letters of Kabbalah (reading backward) correspond to ק = 100, ב = 2, ל = 30 and ה = 5. What does this mean?

6.7 Magnetism and Spin

Electrostatics holds atoms together. Quantum mechanics keeps them from collapsing. But, there is more. Magnetic forces are also significant. Because the average speed of an electron in the hydrogen atom is much smaller than the speed of light (it is roughly $c/137$), one expects the magnetic energies to be much smaller than the 13.6-electron-volt binding energy of hydrogen. The electrons in more massive atoms move faster, so magnetism and special relativity become more important.

Magnetism in the quantum world has two parts. The intuitive part is analogous to classical physics, with the motion of charges producing magnetic fields. This is "orbital" magnetism. The counterintuitive part is called "spin" magnetism. For an electron, spin defies common sense because an electron has no size. Despite its vanishing radius, an electron's spin is the source of both magnetism and angular momentum.

6.7.1 Orbital Magnetism (Ignoring Spin)

The orbital angular momentum and magnetism of an atom are related. An orbiting electron and its associated magnetic field are sketched in Figure 6.17. Two equations for magnetism on a large scale remain valid on the atomic scale:

$$magnetic\ moment = \frac{e}{2m}(angular\ momentum)$$

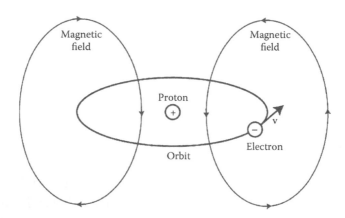

Figure 6.17 An electron circling a nucleus has a quantized angular momentum and a quantized orbital magnetic moment.

and

Magnetic energy = –(Magnetic moment) × (Magnetic field)

The exciting new aspect of atomic magnetism is the quantization of angular momentum. This quantization determines the magnetic moment exactly, so a precise measure of the magnetic energy gives an equally precise value for the magnetic field.

6.7.2 Spin

Quantum Rule 5

Electrons, protons, and neutrons have spin.

Spin is the surprise. An electron's spin angular momentum, like its orbital angular momentum, is quantized. The quantization rule is a little different. The spin can only be "up" or "down." The magnetic moment of the two spin orientations is the same as the orbital magnetic moment when the quantized angular momentum is either $+h/(2\pi)$ or $-h/(2\pi)$. The electron's spin and resulting magnetic field are sketched in Figure 6.18. Actually, there is a small correction to the spin magnetic moment described by quantum electrodynamics (Chapter 10). The fractional increase is close to

$$\frac{1}{2\pi}\frac{1}{137}.$$

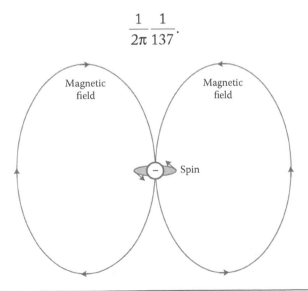

Figure 6.18 This picture of a spin-up electron and its corresponding magnetism is not realistic because it suggests a nonzero radius for the electron.

The proton also has a spin angular momentum. It obeys the same quantum rule that says the spin can be either up or down. The proton also has a magnetic moment, but it is about 658 times smaller than the magnetic moment of the electron. The major difference comes about because the proton is about 1800 times as massive as the electron. There is no simple expression for the proton's magnetic moment because, unlike the electron, the proton has size and structure. It can seem surprising that a neutron, with no charge, also has a magnetic moment. The magnetism is produced by the charged "quarks" that combine to make both the proton and the neutron.

6.7.3 Magnetic Resonance

To say atomic magnetism is useful is an understatement. The quantization of magnetic moments allows us to "see" a hidden world through high-precision measurements of magnetic fields. The key is the energy of a magnetic moment in a magnetic field. Magnetic resonance imaging (MRI) uses the quantized spin of protons in a hydrogen atom to probe the magnetic field. For example, MRI can detect slight magnetic differences between bone and cartilage. If one has a sore knee, MRI could help to diagnose the problem. MRI is also an important tool for cancer detection. Magnetic fields are not dangerous. This is an important advantage of MRI compared with x-ray imaging.

6.8 Many Particles

Quantum mechanics would surely have been more difficult without the hydrogen atom. Its single electron flying about the heavy proton made the physics about as simple as possible. Unraveling the mysteries of the hydrogen atom was the key that culminated in the Schrodinger equation. With the refinement of spin, the hydrogen atom description is quite accurate. But, nature is not just hydrogen atoms. A generalization for many particles is needed to describe the real world. The simplicity of hydrogen is lost even for the helium atom with its two electrons. Heavier atoms, molecules, and condensed matter present even more challenges for quantum mechanics.

Many electrons pose two new questions: How does one deal with identical particles? How does one find a practical method to solve many-body problems? Hints at the answers to these questions follow.

6.8.1 Helium

The next logical atom to consider is the two-electron and two-proton (and two-neutron) helium atom, shown with a greatly expanded nucleus in Figure 6.19. Helium (named for Helios, the Greek sun god) was first observed on the sun in 1868 (during a solar eclipse) by the yellow light that corresponds to the energy of one of helium's atomic transitions. It took 27 more years before helium was isolated on Earth.

As with hydrogen, it is a good approximation to view the small, heavy helium nucleus as a fixed point of attraction for the two electrons. That is the easy part. The hard part is finding the wave functions and energies. The real difficulty with helium is the repulsive Coulomb force between the two electrons. This force pushes the two electrons toward opposite sides of the atom. This annoying "correlation" means the Schrodinger equation cannot be solved exactly. One must rely on clever tricks and the help of computers to obtain the right answers. It should not be surprising that helium is so difficult. Even the classical three-body problem of two identical planets circling a star has no easy exact solution.

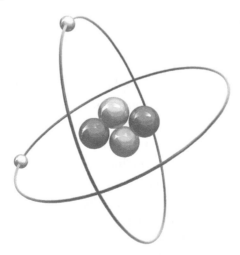

Figure 6.19 Unrealistic helium atom with two electrons. Electrons should be described by waves. The nucleus with two protons and two neutrons is much smaller than depicted.

6.8.2 Identical Electrons

There is no way to tell one electron from another as they are identical, and this ambiguity must be reflected in the wave functions that describe more than one electron. The result is a rule.

Quantum Rule 6

A multielectron wave function must change sign whenever two electrons are interchanged. The same rule applies to protons and neutrons.

Quantum rule 6 is plausible because no measurement should change if indistinguishable particles are interchanged. The probability for finding any result is determined by the square of a wave function, so probabilities will not change if the wave function only changes sign. (The square of a negative wave function is the same as the square of the original wave function.)

One can argue that the sign change is unnecessary for preserving the probability. It would be simpler to require no sign change with particle interchange. The clever part of the reasoning (proved by Pauli) is that the sign change is required for identical particles that have spin-up and spin-down orientations. This sign change rule applies to electrons, protons, and neutrons because all these particles have the same spin property. Photons are different. There is no sign change when two photons are interchanged.

6.8.3 Pauli Exclusion Principle

No two electrons can be in the same quantum state.

This exclusion principle is an application of quantum rule 6. The simplest example is two-electron helium. Assume each of the two electrons is described by its own wave function. If the electrons were described by exactly the same wave function and if the two electrons had the same spin direction, an interchange would leave the wave function unchanged. This is not good. The wave function must change sign. This means the electrons cannot be in the same *quantum state*. (Quantum state means wave function and spin direction.) For helium in its ground state, the electrons are put in different states by making one of them "spin up" and the other "spin down." The opposite spin directions mean the helium atom has no magnetism.

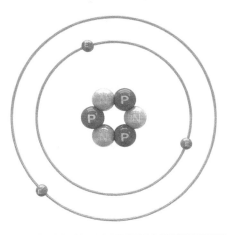

Figure 6.20 Two electrons with opposite spin can be found near the lithium atom nucleus. The Pauli exclusion principle requires the third electron to lie further out (on average). The figure unrealistically makes it appear that the identical electrons can be distinguished.

The antisymmetry requirement for identical electrons is exact. Pauli's exclusion principle is not. Many-electron wave functions are generally complicated. Two electrons cannot be described by two single-electron wave functions, so one cannot really say each electron has its own state or its own energy. Despite this, the exclusion principle combined with imaginative approximations yields abundant useful results.

The lithium atom of Figure 6.20 has three electrons. That means two of the electrons in lithium must have parallel spins. (Up and down are the only spin choices.) The presence of three electrons and the Pauli principle require that a wave function of a different shape describe one of the electrons. An electron described by this second wave is typically further from the nucleus. In practice, this means relatively small energy is needed to remove an electron from a lithium atom. In the language of an approximate "shell model," one says the third electron must be placed in an outer shell because the first two electrons "fill" the inner shell. Extended applications of ideas related to the shell model have allowed people to develop approximations that adequately describe atoms, molecules, and more complicated systems. Despite some complications associated with the approximations of the Pauli principle, Mendeleev's periodic table can, at last, be explained.

7

MATERIALS AND DEVICES

7.1 Introduction

One need not apologize when considering the applications of physics to down-to-Earth objects.

Physics rightly deals with both formal foundations (like supersymmetry) and the properties of materials (like superconductors). Supersymmetry and superconductors do not appear to have much in common. However, the surprising overlap between abstract ideas and their applications to real materials can be rewarding. For example, the generalizations of ordinary phase transition (like melting) are seen in theoretical particle physics. The applications of scale invariance (big looks like small) have helped to unify material physics, particle physics, and cosmology.

Physics is supposed to describe everything. Thus, one might expect a physicist sufficiently skilled in classical and quantum physics to know everything about solids, liquids, and gases; devices; chemistry; and even biology. Of course, physicists do not know all because the world is too complicated. However, basic physics can explain a great deal about the materials of our everyday world, and physics has contributed to the development of the devices (like toasters and computers) that we take for granted.

The physics of materials is often a story of experiments leading theory. The ancient inventors of bronze did not follow a theoretical lead. There was no theory. The twentieth-century discoveries of superfluids and superconductors were a complete surprise. The DNA model was proposed only after x-ray experiments revealed its geometry. On the other hand, calculations describing properties of graphene (a single layer of graphite) were done long before application of the trick (the famous Scotch® Tape technique), which enabled separation of single layers of carbon atoms.

Many useful devices were developed with the help of a solid theoretical background. Edison certainly understood the basics of sound when designing his phonograph. More recent developments, such as lasers, transistors, and computers, evolved as theory and experiment worked together.

7.2 Materials

This brief and restricted description of materials only hints at the enormous variety of useful and exotic substances known today. No doubt, discoveries of new materials and new applications of old materials are just around the corner.

7.2.1 Solid, Liquid, Gas

Ice, water, and steam are the three phases (solid, liquid, gas) of H_2O. Most materials exist in these three phases, but there are ambiguities and exceptions. For example, many different ice crystal structures can be produced at high pressures. Thankfully, none of these has the properties of "ice-nine," a mythical substance conceived by Irving Langmuir and elevated to the role of a central plot device in Kurt Vonnegut's novel *Cat's Cradle*. In the novel, ice-nine froze at room temperature. Once a crystal started, it quickly spread and solidified all water on Earth. Everyone died. Langmuir's ice-nine fantasy suggests creativity worthy of the Nobel Prize (in chemistry) that he received in 1932.

7.2.1.1 Attraction: Key to Condensation   Materials do not remain gases composed of free-flying atoms at low temperatures. Solids and liquids are formed by an attraction between atoms. The details of the force involve the interplay of electrostatics and quantum mechanics. The different forces for different atoms explain the variety of structures and properties of solids and liquids.

Some common features of the forces that lead to condensed matter (solid or liquid) can be seen by considering the force between just two atoms. At large separation, there is an attraction even between neutral atoms because the electrons in each atom perform a dance that allows them to share the attraction from the neighboring atom's nucleus. When the positive nuclei get too close, they push on each other. Atoms pull together until attraction and repulsion balance. This ideal separation is roughly the sum of the two approximate atomic radii. The simplest hydrogen molecule example is described in Chapter 6, Figure 6.15.

7.2.1.2 Solid or Liquid   The diversity of materials means generalizations have exceptions. Even the solid, liquid, and gas categories can be confusing.

Ideally, solids are distinguished from liquids and gases by their ordered crystal structure. There is no regularity in the atomic positions in liquids or gases. But, very hard glass has no crystal structure. Its atoms appear to be about as random as the atoms in a liquid (see the discussion and figure in Section 7.2.4.2). On the other hand, a "liquid crystal" flows like a liquid even though its long molecules remain oriented along a fixed direction. An LCD (liquid crystal display) uses the anisotropic interaction of the long-oriented molecules with light to produce the images seen on many televisions, computers, and other electronic devices, as described in Chapter 5 (Section 5.5.4).

7.2.1.3 Liquid or Gas   The distinction between a liquid and a gas can be equally confusing. One conventionally thinks that gas atoms fly about freely and only occasionally collide, while the atoms in a liquid are nearly as compact as in solids. But, when a gas is subjected to increasing pressure, the atoms are pushed together. At a sufficiently high pressure, the distinction between a gas and a liquid disappears. Then, a single word—*fluid*—denotes this general state.

7.2.2 Crystalline Solids

At a sufficiently low temperature and atmospheric pressure, all but one of the elements solidifies into a periodic crystalline solid. The exception is helium, which only solidifies under high pressure. Many compounds also crystallize. Snowflakes, table salt, and quartz are familiar crystals.

The visible symmetries associated with crystals hint at their periodic structures. Indeed, long before atoms or molecules were known,

Johannes Kepler speculated in 1611 that the sixfold symmetry of a snowflake was related to the periodic arrangement of tiny water droplets. (The most compact way to place pennies on the table is a hexagonal lattice with sixfold symmetry.) Kepler was probably not the first, and surely not the last, to speculate on the possible periodic underpinning of crystal symmetries.

Today, x-rays are the experimental workhorse of crystallography. The underlying physics is a generalization of Young's double-slit experiment. When x-rays scatter from planes of atoms, the waves add to produce maximum intensity when the scattering distance from different crystal planes differs by an integer number of wavelengths. Determining the distance between various crystal planes can reveal the atomic positions. An x-ray is the proper choice of electromagnetic radiation because its wavelength is comparable to the interatomic spacing.

A fundamental discovery of biology came about, in part, from the x-ray determination of a crystal structure. The "double-helix" model of DNA that is the foundation of genetics (and of life) was presented by James Watson and Francis Crick in 1953. The x-ray studies of a crystallized form of DNA were crucial for their work. Rosalind Franklin played an essential role in this research. Like many fundamental breakthroughs in science, the history of DNA and the determination of its function and structure have a complicated and controversial history. Suffice it to say that the discovery of the double helix was the culmination of ideas from many people over a number of years. A very small piece of DNA is idealized in Figure 7.1.

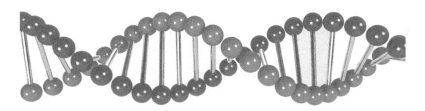

Figure 7.1 A simplified schematic view of a portion of the DNA molecule whose structure can be obtained using x-rays.

7.2.3 Three Important Crystals: Salt, Diamond, and Copper

7.2.3.1 Salt   Table salt, NaCl, looks like Figure 7.2a and has the cubic crystal structure shown in Figure 7.2b. Deducing its relatively simple structure is an easy job for x-rays.

One can roughly "explain" the stable structure of salt by looking at the electric properties and sizes of the constituent atoms. Sodium atoms have a less-firm grip on their electrons than the chlorine atoms, so they end up with a positive charge and the chlorines become negative. The NaCl crystal structure minimizes the electrostatic energy by placing each positive sodium atom in a cage of six negative chlorines. There are alternative crystal structures that can minimize the electrostatic energy, but the cubic structure of Figure 7.2b is especially favorable for NaCl because the smaller sodium atoms (having lost electronic charge) fit snugly into the holes of the chlorine atomic cages.

Salt is rigid because atoms are firmly locked in position. A sodium atom does not want to move to a neighboring site because of its comfortable position in the chlorine cage. Salt is also brittle because the motion of one sodium weakens the neighboring structure, resulting in a propagating crack.

7.2.3.2 Diamond   Diamond, made entirely of carbon atoms, is harder than salt (and everything else). A diamond and a bit of its crystal structure are shown in Figure 7.3. Chemistry suggests the stability

(a)   (b)

Figure 7.2   (a) Salt. (b) A salt crystal. The sodium atoms, having lost electronic charge, are slightly smaller than the chlorines.

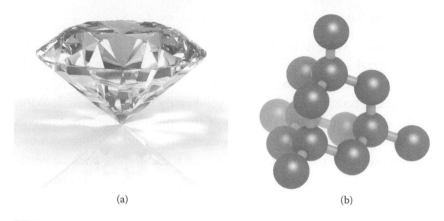

(a)   (b)

Figure 7.3   (a) A diamond. (b) A small piece of diamond lattice showing that each carbon atom has four nearest neighbors.

of the diamond crystal. Each carbon atom is particularly stable when it forms strong chemical bonds with four other carbon atoms. These rigid bonds give diamond its remarkable properties.

A diamond is really not "forever." At atmospheric pressure and temperature, the hexagonal graphite form of carbon is more stable than diamond. Eventually (meaning a really long time), a diamond will turn into commonplace graphite. This transformation can be speeded up enormously if one makes the mistake of throwing a diamond into a fire, where it will turn to graphite or burn and become carbon dioxide.

Diamonds are rare because of their instability on Earth's surface. They are formed where they are stable, deep underground at high temperature and pressure. We find natural diamonds only when they are carried upward by geologic upheavals, like volcanoes.

Laboratories can produce the high temperatures and pressures needed to manufacture diamonds. Some of Ludwig von Beethoven's hair (with some additional carbon) was turned into a diamond. There are companies that can change a beloved pet (deceased and cremated) into a diamond—at a price. Tequila has also been changed to diamond. Because diamond is essentially a simple carbon lattice, it is extremely difficult for a casual observer to distinguish real and manufactured diamonds. Only imperfections and impurities give a diamond its color and character. The De Beers cartel has developed special optical instruments designed to detect manufactured diamonds. It is not clear that the people who manufacture diamonds are interested in flooding the market and reducing diamond prices.

7.2.3.3 Copper   Copper atoms are not held together by the strong electrostatics of NaCl or the rigid bonds of diamond. Instead, copper (Figure 7.4a) has a more democratic approach. Each atom wants to be near as many neighbors as possible. The most stable structure centers each copper within a shell of 12 nearest neighbors. This compact atomic arranging is accomplished by a face-centered-cubic lattice, a piece of which is shown in Figure 7.4b. Copper is softer than salt or diamond because the attraction between copper atoms is neither rigid nor directional.

7.2.4 Imperfect Crystals and Disordered Solids

All crystals contain impurities and defects, so the ordered structure is only an approximation. Sometimes, even small impurity concentrations

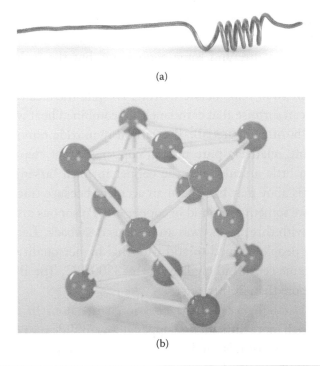

(a)

(b)

Figure 7.4   (a) A piece of copper wire. (b) A single cube of the face-centered-cubic lattice showing an atom at each face center.

make a big difference. Pure diamond is transparent for all visible wavelengths and shows no color. One boron atom for every million carbon atoms gives the crystal a bluish tint. The right type of imperfections in a diamond can make a gem much more valuable.

Impurities in copper reduce its electrical conductivity. This is especially noticeable at low temperatures, at which impurity scattering can be the major source of the electrical resistance. Some have attributed impurities in salt with special health benefits. Sea salt is an example, but there is no scientific evidence that the additional benign salts of potassium, magnesium, and calcium as well as some sulfates will do anyone any good.

7.2.4.1 Bronze Bronze is tough. The Trojan War (assuming one believes it took place) was probably fought with weapons of the late Bronze Age. Bronze is typically copper with about 10% tin. The added tin atoms mean the crystal structure is complicated and only approximately periodic.

Copper and bronze played a central role in the development of civilization. People learned how to obtain copper from copper ore before 5000 BC. Initial refinement may have been an accident associated with a hot fire and some copper ore, but the technique was refined over the ages. It is remarkable that fairly difficult metallurgy was invented 2000 years before writing. Despite the lack of a written record, we can surmise that early smelting combined heat with carbon monoxide (burning charcoal). The carbon monoxide removed impurities, leaving relatively pure molten copper. A first step to bronze combined a little arsenic with the copper to form "arsenic bronze." Initially, this alloy may have been an accident because trace amounts of arsenic are sometimes found in copper ore. Poisonous arsenic vapor associated with this process was an obvious drawback. Later (around 4000 BC), people learned to add tin. This higher-quality and safer bronze became central to civilization by 3300 BC. The Bronze Age lasted for more than 2000 years.

Bronze is tough because the occasional tin atom in the crystal makes it much more difficult for copper atoms to slide past each other. Imperfections in crystals can be introduced other ways, such as "work hardening," which juggles atomic positions and reduces the ordering, resulting in hardness. Hard materials are often brittle, so there is a compromise. Sometimes, partial ordering can be restored with annealing or heating to an intermediate temperature.

Bronze is but one of many examples of important alloys. The properties of steel depend on the amount of carbon and other materials mixed with the iron. Additions of molybdenum and samarium make steel stronger at high temperatures. This was of military importance for cannons like Germany's "Big Bertha" of World War I.

7.2.4.2 Glass and Amorphous Solids   Is glass a solid or a liquid? This simple-sounding question does not have a simple answer. Glass is hard, so it is solid. But, glass has no crystal structure, so it is liquid.

Glass and ice melt differently. When ice melts it exhibits a clear "phase transition." A large amount of heat must be added to change ice to water. That is why ice cubes do a much better job of cooling a drink than an equal amount of cold water at almost the same temperature.

As glass is heated, a gradual transition from solid to liquid takes place as it becomes increasingly soft. It is silly to identify an exact temperature at which melting takes place. However, at the extremes, the difference is clear. The molecules in cold glass can move only short distances. The molecules in molten glass can move past each other with relative ease.

The clear distinction between a crystalline solid and a glass (or liquid) is illustrated in Figure 7.5. Atoms are ordered in idealized crystal at the left, but disordered in the idealized glass on the right.

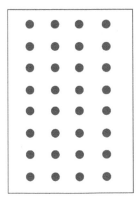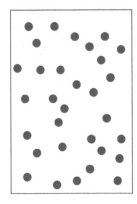

Figure 7.5   The atomic positions on the left clearly identify a crystal solid. The positions on the right could be a gas, a liquid, or an amorphous solid. One can identify the difference only by watching atomic motion.

The materials we commonly recognize as glass are composed of silicon dioxide molecules and smaller amounts of various other materials (often other oxides). Many familiar materials other than glass show a similar lack of crystal structure and gradual melting. These materials are called *amorphous*. Amorphous examples include most plastics.

7.2.5 Metals and Insulators

Stated most simply, metals conduct electricity; insulators do not. The difference between a metal and an insulator can be quite striking. *Conductivity* gives a quantitative measure of the distinction. The conductivity (measured in inverse ohm centimeters) is the amount of electrical current flowing through a 1-centimeter cube of material when a 1-volt difference is applied to opposite faces of the cube. For pure copper, that current is more than half a million amps. Diamond is an example of the other extreme. The conductivity of the purest diamond is almost immeasurably small. Even with a few impurities, applying the 1-volt difference to a diamond cube will produce far less than a billionth of an amp.

Copper differs from diamond because electrons are free to move about the copper lattice. Copper's extremely mobile electrons make it the material of choice for the delivery of electricity to everyone's residence. The copper electrons also carry energy, so it is a good heat conductor. Generally, metals are the best thermal conductors, but diamond is a striking exception. It conducts heat better than any other material because its rigid lattice vibrations carry energy almost perfectly.

Many materials are not simply classified as insulators or conductors. The semiconductors have electrical conductivities that vary greatly with impurity content and temperature. Silicon is a semiconductor that has the same lattice structure as diamond. However, its less tightly bound electrons can be coaxed into participating in electrical conduction. Semiconductors are the foundation of many modern electronic devices. Some of their applications are described in Section 7.3.

7.2.6 Special Properties

7.2.6.1 Magnets
A ferromagnet spontaneously generates a magnetic field. Iron is probably the best known of many examples that include

nickel and cobalt. Ferromagnetism disappears when a substance is heated above a transition temperature (1043 K for iron). Many refrigerator doors are adorned with ferromagnetic permanent magnets. Even though these magnets are familiar objects, their physics is not simple. Theories can explain the generalities of ferromagnetism, but not easily and not with the precision needed to predict the transition temperature. Ferromagnetism is intrinsically mysterious because classical theory tells us that magnetism can only be generated by electric currents. The electron spin and quantum mechanics of materials must be considered to properly understand ferromagnetism.

7.2.6.2 Reduced Dimensions   Three notable examples of reduced dimensional materials are made of carbon atoms. They are graphene, carbon nanotubes, and buckyballs. Graphene, shown in Figure 7.6, is a single layer of carbon atoms in the form of a honeycomb lattice. Its remarkable properties suggest it may soon find exciting applications. If many layers of graphene are placed on top of each other, the graphene layers become ordinary graphite.

A carbon nanotube (Figure 7.7) is a graphene lattice curled up to make a tube. Buckyballs (Figure 7.8) are roughly a spherical form of this structure. The simplest buckyball is made of 60 atoms that lie at points described by the geometry of a soccer ball.

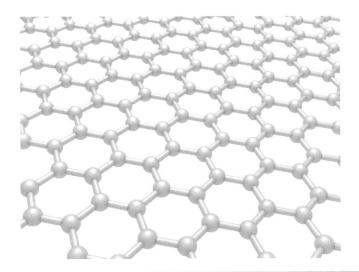

Figure 7.6   A piece of the graphene lattice.

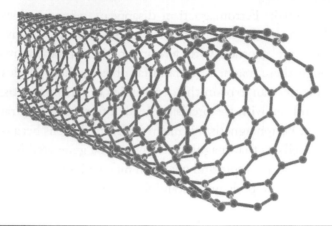

Figure 7.7   Carbon nanotube.

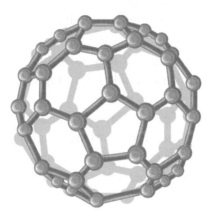

Figure 7.8   Buckyball of 60 carbon atoms.

7.2.6.3 Plasmas   A plasma is sometimes called the fourth state of matter. Plasmas are gases associated with temperatures so high that many of the electrons are no longer bound to their nuclei. The sun is plasma. Attempts to produce controlled fusion using magnetic confinement use plasmas.

Plasmas are the basis of fluorescent lights. For these lights, some of the electrons are removed from atoms of a dilute gas by an electric field rather than heat. Energy is released when an electron recombines with one of the ions. This energy is often a photon. Energy conservation determines the light frequency through the relation *Energy* = *hf*. As always, *h* is Planck's constant, and *f* is the photon frequency. This recombination energy is often so large that only invisible ultraviolet light is produced. A fluorescence material coating the surface absorbs the ultraviolet photon and emits lower-frequency light. Most of the energy left over in this photon trade becomes heat.

Plasma televisions are a remarkable application of this light source. Each pixel of the television screen contains three very small red, green, and blue plasma lights. When these pixels work in concert, an impressive image is produced.

7.2.6.4 Superconductors When Kamerlingh Onnes discovered super-conductivity in 1911, he first thought there must be some mistake. When the electrical resistance of his mercury sample completely vanished at 4 K (very cold), he looked for a short circuit. He checked, but there was no mistake. He had discovered a new state of matter with unusual properties.

Superconductors have other properties besides infinite conductivity. They repel magnetic fields, so a superconductor can be levitated by a magnet. Some superconductors circumvent the "no magnetism" rule and allow magnetic fields to penetrate as an array of quantized magnetic flux bundles. Quantum effects associated with superconductors mean devices can be made that detect very small magnetic fields. Many metals become superconducting at low temperatures, and the vanishing resistance has important applications, especially for producing large magnetic fields. The fields produced by superconducting magnets are used in high-energy physics to steer energetic particles. They are also employed in magnetic resonance machines, which are an increasingly important tool in medicine.

Sometimes, it takes a long time to explain a new discovery. For superconductivity, the successful theory of Bardeen, Cooper, and Schrieffer did not appear until 1957, which was 46 years after Onnes's original discovery.

A new class of "high-temperature superconductors" appeared with the first discovery of Bednorz and Muller in 1986. For some of these materials, the transition temperature is above 100 K. This is still pretty cold by ordinary standards, but the potential applications are impressive. The theory of this new form of superconductivity appears to be complicated, involving significant modifications of the basic theory of Bardeen, Cooper, and Schrieffer.

7.2.6.5 Superfluids Superfluidity may have saved the life of Lev Landau, a brilliant Soviet theorist. Landau was jailed in 1938 for unwisely comparing the Stalinist dictatorship with that of Hitler. A year earlier, superfluid helium was codiscovered by another Soviet scientist, Pyotr Kapitsa. Kapitsa appealed for a very ill Landau's release, with the argument that only Landau could explain helium's many mysterious properties. Kapitsa was successful, and he was correct. Landau was released, and he later won the 1962 Nobel Prize for his theory of superfluid helium.

Superfluid helium has no viscosity. It is analogous to superconductivity because fluid can flow with no applied pressure, just as an electric current flows through a superconductor with no applied voltage. Helium becomes a superfluid only below 2.17 K (really cold). Heat propagates through a superfluid like a sound wave. Superfluid helium can creep over the wall of its container. Rotating buckets of superfluid helium reveal angular momentum quantization on a macroscopic scale.

One key to understanding both superfluidity and superconductivity lies in the symmetry or antisymmetry of many-particle wave functions. The wave function for helium atoms is symmetric when two atoms are interchanged. The six sign changes resulting from the interchange of two electrons, two protons, and two neutrons are equivalent to no sign change. There is a second "isotope" of helium that has only one neutron. The one-neutron helium atoms are antisymmetric on interchange, and this rare version of helium does not become a superfluid at 2.17 K. Symmetric wave functions are also associated with superconductivity. The zero-resistance current is carried by "Cooper pairs" of electrons. Paired electron wave functions are also symmetric on interchange.

7.3 Devices and Applications

7.3.1 Microwave Ovens

These handy household microwave oven devices generate electromagnetic waves with a frequency of 2.45 billion hertz and a wavelength of about 12.2 centimeters. The waves heat our hot dogs, but they do not heat the glass plate.

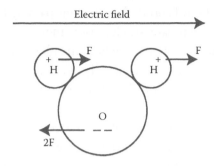

Figure 7.9   Water molecule. Small positive charges on the hydrogen atoms and a negative charge on the oxygen mean the electric field exerts a torque on the molecule.

7.3.1.1 Heating Water   The ability of microwaves to heat food lies in a special property of the water molecule shown in Figure 7.9. The bent shape of the water molecules combined with small charges (positive on the hydrogen and negative on the oxygen) mean an electric field will exert forces on the atoms and a torque on the molecule.

The twisting produced by the electric field energizes the water molecules, and molecular collisions spread this rotational energy and heat the food. Most molecules do not have water's special properties, but this is not a problem. Meats are around 75% water. Carrots are 90%.

7.3.1.2 Making Microwaves   The magnetron is the standard device used to generate microwaves. It uses heat, electric fields, magnetic fields, and resonance to create the radiation, as sketched in Figure 7.10.

Electrons are accelerated from the heated central region of the magnetron by an electric field. A magnetic field changes the electron

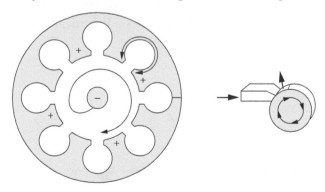

Figure 7.10   An end view of the cylindrical magnetron geometry and the analogous geometry of a whistle.

trajectory to an outward spiral. As electrons pass by the cavities in the outer section, they generate an electromagnetic resonance similar to the acoustic resonance generated when one blows into a whistle.

Microwave ovens are efficient. Typically, 1100 watts of electric power are converted into 700 watts of microwaves. Most of the remaining energy is lost as heat in the magnetron.

Microwave energy penetrates about 1.7 centimeters into the food. The energy deposited is largest at the surface and decreases with depth as the waves become weaker. It is not true that microwaves cook the food from the inside out.

7.3.2 Lasers

Laser light is different from ordinary light. This single-frequency light travels in a precisely defined direction. The photons that make up laser light move in concert, so the waves associated with the photons add together. Symmetry again plays a role. Photon wave functions are symmetric when two are interchanged. The exchange symmetry increases the chance that two photons will be described by the same wave. Idealized laser light is many photons all described by the same wave.

A photon is produced when an atom makes a transition from a higher-energy level to a lower level. This is illustrated in Figure 7.11. The atom in its high-energy level is on the left, and the atom in its low-energy level along with an emitted photon are on the right. Energy conservation means the photon energy equals the difference between the two atomic energies.

Time reversal symmetry applies to classical physics and to quantum mechanics (with some technical quibbles). This means the process in

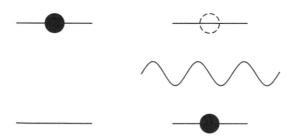

Figure 7.11   At left is an atom in a higher-energy level. At the right, the atom in the lower-energy level is accompanied by an energy-conserving photon.

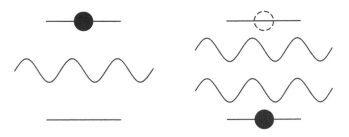

Figure 7.12   With an additional photon present, the rates for both the absorption and the radiation are doubled.

Figure 7.11 can occur from left to right or from right to left, so photon emission and absorption can both occur.

The rate at which a photon is absorbed to lift the atom to the higher-energy level is proportional to the number of photons. So, if a second photon is present, as is shown on the right of Figure 7.12, the absorption rate is doubled. The time reversal symmetry means that the radiation rate in the presence of a single photon is also doubled. Photons stimulate the radiation. This rule generalizes, and it is the key to laser action.

A laser operates by placing a majority of the atoms in the higher-energy level. Then, when one atom makes the transition to the lower-energy level, the radiated photon stimulates additional radiation. The process can quickly "snowball" because many atoms are stimulated to drop to the lower level and produce many essentially identical photons that are the characteristic of laser light.

Coaxing more than half the atoms into the upper level is a necessity for laser light. If less than half the atoms are in the upper level, photon absorption would overwhelm the emission, and the laser will fail.

Heat cannot make a laser. No matter how high the temperature, the number of atoms in the higher-energy level is always less than the number of atoms in the in the lower level. Heat does not maximize energy. It makes more energy available so higher-energy levels are nearly as likely, but the higher-energy levels are never preferred.

A variety of methods can prepare a system so that more atoms are in the higher of two energy levels. High-frequency light, atomic collisions, and the injection of energetic electrons (light-emitting diode or LED) are all capable of producing the "population inversion" necessary for laser operation. For the LED, the inversion is characterized by more electrons in energetic states that are not necessarily tied to atoms.

Lasers have applications ranging from erasing tattoos to bouncing light off the moon. This once-academic curiosity has become a standard tool with an ever-increasing list of uses.

7.3.3 Semiconductors and the Amazing p–n Junction

7.3.3.1 Semiconductors Not all materials are simply classified as metals or insulators. Semiconductors are a more complicated case with many technological applications. A pure semiconductor, like silicon, is an insulator. However, even a small number of properly chosen impurities added to a semiconductor can greatly change its electrical conductivity.

A "donor" impurity has a loosely bound electron. This electron escapes from the donor and hops about the semiconductor crystal. The freed electron produces significant electrical conductivity, and the semiconductor is no longer an insulator. Because negative electrons carry the electrical current, these are called *n-type* semiconductors.

Although it is a little harder to visualize, one can also form *p-type* semiconductors in which the current appears to be carried by a positive charge. The charge is called a *hole* because it is actually a missing electron hopping from site to site. *P*-type semiconductors are produced by adding "acceptors" that grab electrons from the semiconductor atoms.

7.3.3.2 The p–n Junction A p–n junction is obtained by adding "donors" and acceptors to adjacent pieces of a semiconductor. Simplified illustrations of *n*-type and *p*-type semiconductors and a p–n junction are shown in Figure 7.13.

One might think the positive and negative charges of a p–n sandwich would combine to make neutral objects, leaving no charge to carry current. After all, electrons and protons produce neutral and nonconducting hydrogen atoms. Neutralization does not occur except near the p–n junction because the mobile electrons in the *n*-type material swim in a background of positively charged donor ions. Similarly, the mobile holes are electrostatically compensated by the negatively charged acceptor ions. When an electron combines with a hole, the mobile charge disappears, but the background charges of the acceptor

Semiconductors

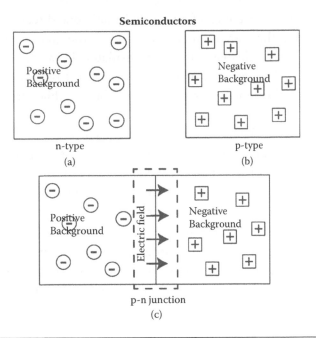

Figure 7.13 Top left, an *n*-type semiconductor; top right, a *p*-type semiconductor; bottom, a p–n junction.

and donor ions remain. Because the background ions cannot move, the neutralization of mobile electrons and holes leaves net charges behind in the p–n junction where the mobile charges combine. This background charge is positive in the n-type region and negative in the p-type region.

Adjacent regions of positive and negative charge produce an electric field directed across the p–n boundary. This field is a microscopic analogy of the capacitor used in Millikan's oil drop experiment. The electric field at the p–n junction pushes both the positive and negative mobile charges out of the junction, preventing further neutralization of electrons and holes.

The p–n junction leads to a number of applications. Some are listed here.

1. A diode that allows current to flow in only one direction
2. A light-emitting diode
3. A semiconductor laser
4. A solar cell

7.3.3.3 Diode   Current can flow easily in only one direction across a p–n junction. There are no mobile charges in the junction region, so one might expect this layer to be an insulator. However, an electric field or applied voltage can produce a current if it pushes the mobile charges into the junction region (called forward bias) (see Section 7.3.4.1). As the electrons and holes cross the junction, they combine as they meet, but the net effect is an electric current.

It is not possible to produce electric current in the opposite direction. Pulling positive and negative charges away from the p–n junction (called reverse bias) increases the electric field and enlarges the region with no mobile charge carriers.

Diodes play many important roles in electronic circuits. One example is the changing of an alternating current to a direct current. A diode can eliminate the negative part of the sine wave that characterizes alternating current. Capacitors can smooth out the resulting upper half of the sine wave.

7.3.3.4 Light-Emitting Diode   It typically takes a few volts of forward bias to push the electrons and holes into the barrier region of a p–n junction. This voltage supplies an energy that is released as the electrons combine with the holes. For carefully engineered semiconductors, a large part of this energy is converted into photons (light). This LED can convert electrical energy to light more efficiently than an ordinary incandescent lightbulb. Examples are shown in Figure 7.14. A huge

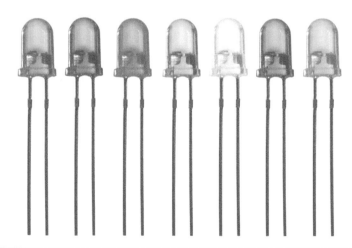

Figure 7.14   Light-emitting diodes. Different semiconductor properties produce different colors.

array of very small LEDs is used for the display screens of televisions and other devices. The technology of LEDs is rapidly progressing.

7.3.3.5 Semiconductor Laser Laser action can occur when there is "population inversion" and high-energy states are more abundant than low-energy states. For semiconductor lasers, this population inversion is produced in the p–n junction, where electrons and holes are pushed together with the help of a forward bias. Considerable engineering is needed to make sure the produced light is sufficiently trapped so that a cascade of stimulated transitions can take place.

Today, semiconductor lasers are the most common lasers. We do not notice them because they are usually tiny. For example, every compact disc player uses a semiconductor laser. The laser light is reflected from the disc's surface, and variations in the reflection are translated into a digital signal.

7.3.3.6 Solar Cells and Digital Cameras A solar cell is an LED operating in reverse. When light is applied to the p–n junction, the photon energy is changed to liberated electron-hole pairs. These charges produce a current in the diode because they are pushed by the p–n junction's electric field. The product of the current and the natural diode voltage is electrical power. Because the solar cell converts light directly into electrical energy, there is no thermodynamic limit on the efficiency of this "photovoltaic" process. However, practical problems mean solar cells are far from 100% efficient. Efficiency continues to improve, and fabrication costs are dropping.

Solar cells are, in theory, and ideal energy source. Using the sun to produce electricity seems almost too easy. But, there are limitations associated with efficiency, cost, lifetime, and energy storage. Intensive research is being devoted to these problems. Some predict that photovoltaic sources will provide a significant contribution to the world's total energy consumption within a decade or two. Predicting the future is not easy. Developments of basic science, engineering, and politics will determine the large-scale implementation of photovoltaics.

Digital cameras also use light to move electrons. Light from different places is focused on different pixels in the back of the camera. Photons release trapped electrons. This results in a charge that is related to the amount of light that arrived at the particular pixel.

Additional (and complicated) electronics are needed to "read" the charge on each pixel after the image is produced.

7.3.4 Transistors

7.3.4.1 Transistor A transistor uses the variable conductivity of a p–n junction to control a current. If the p–n junction of Figure 7.15 is biased in the forward direction by the application of a positive voltage on the right and a negative voltage on the left, charge carriers are injected into the junction. If a second voltage difference is applied between the top and the bottom of the junction, a transverse current will flow. As soon as the forward bias is removed, the current stops. This is the oversimplified essence of a transistor.

It is hard to overstate the utility of the transistor. It allows one electrical signal to control another. Thus, the transistor can be used as a switch by turning the transverse current off and on. It can also be used to amplify a signal because the transverse current can be much larger than the forward bias current. One truly remarkable property of modern transistors is their miniaturization. Millions of transistors can be placed on a surface only a few centimeters across. The very small size is a major reason that transistors do their electrical work so quickly. A microsecond is a long time for a modern transistor. Gordon Moore made a "law" about the progress of miniaturization that cannot possibly hold true for all time. However, Moore's corollary: "The number of people that site Moore's law doubles every 18 months" appears to be true forever.

The designs developed for transistors and related electronic devices are ever expanding. In particular, there are alternatives to the p–n junction. An especially important example is a sandwich of metal, insulator, and a doped semiconductor, called MOS for metal-oxide semiconductor.

Figure 7.15 is oversimplified in many ways, and it fails to suggest the different types of transistors that can be manufactured and the various junction designs. Transistors can regulate voltages or currents through the introduction of a current, a voltage, or a charge. Working together, transistors can do complex mathematics. A simple example is given in the computer description of Section 7.3.5.

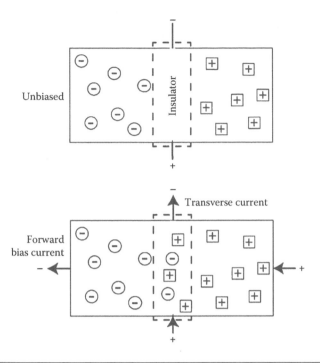

Figure 7.15   The unbiased p–n junction is insulating. Forward bias adds charge to the junction, and it becomes conducting, allowing a forward bias current. With forward bias, the junction can support a second transverse current.

7.3.4.2 Memory Elements   A properly designed transistor forms the basic solid-state memory element. A charge rather than a current produces the forward bias. When the charge is present, the transistor allows current to flow, so a conduction test "reads" the memory by determining whether the charge is present. The charge must be insulated from its surroundings so the memory lasts a long time. Additional tricks are needed to place and remove the charge.

7.3.4.3 Amplifier   A plumbing analogy illustrates the principles of an amplifier and the associated feedback mechanism. The geometry of Figure 7.16 shows that the flow in a stream is being plugged by a submerged bucket. A pump can remove water from a bucket. As the water is removed, the bucket's plug floats up and allows increased stream flow. However, there is a "feedback loop." One-fourth of the water that flows through the stream circles back and is dumped into

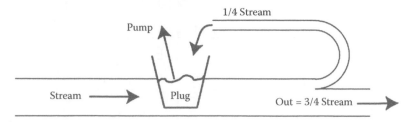

Figure 7.16 A fluid analogy of an electronic amplifier.

the bucket. This slows its rise. Soon, a steady state is achieved in which the water level in the bucket does not change.

The equality of flow in and out of the bucket means

$$Pump = \frac{1}{4} \, Stream$$

From this, it follows that the output (three-quarters of the stream) is

$$Out = 3Pump$$

The sketch of Figure 7.16 is a fluid-flow amplifier. No matter how gently or violently water is pumped from the bucket, the stream output is triple the pump rate. The amplification is 3.

The amplification of this model could be easily changed by altering the geometry. If only 1 part in 11 of the stream flow was returned to the bucket, the flow of the pump would be amplified by a factor of 10.

Amplification of electronic signals is analogous to this plumbing example. The pump flow corresponds to a transistor's forward bias current, and the stream flow corresponds to the transverse current of Figure 7.15. The division of the current into an output component and a feedback component can be done using resistors because Ohm's law tells us that a current is inversely proportional to the resistance.

7.3.5 Computers

The toy computer described here can add two input numbers (both either 0 or 1). If the sum of the two numbers is 2, a light turns on. For any other sum, nothing happens.

The rudimentary computer operation consists of three steps. First, the input numbers are stored. Then, these numbers are read and deliv-

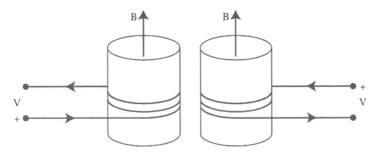

Figure 7.17 The currents magnetize the two bits in the "up" direction, corresponding to the number 1 stored on each bit.

ered to a central processing unit (CPU). Third, the CPU produces an electric current and a light only if the sum is 2.

The numbers are stored in a mini-"hard drive" suggested by Figure 7.17. This hard drive has a capacity of 2 bits. Each bit is a small magnet that is polarized in the "up" direction if the number is 1 and in the "down" direction if the number is 0. A loop of wire surrounds each of the two magnetic bits. The numbers are entered into the toy computer by sending a current through the two wire loops. The currents are sufficient to magnetize the bits. For the example shown in Figure 7.17, each of the input numbers is 1, so both magnetizations are "up."

The numbers in the 2-bit hard drive are retrieved using a circuit containing a material that conducts electricity more easily in a magnetic field. A multilayer sandwich of iron and chromium has this property, called *giant magnetoresistance*. The layered box-shaped objects in Figure 7.18 denote these magnetic field sensors. The sensors are part of a circuit, and there are additional magnets arranged, so a larger current (or voltage) is passed through the sensors when the bit is oriented in the up direction.

The two signals generated by the circuits in Figure 7.18 are sent to the transistor, as idealized in Figure 7.19. When the spin is up in the left-hand bit (#1), the switch is turned "on" by a forward bias on the transistor. When the spin is up in the right-hand bit (#2), a current is delivered to flow through the closed switch. Both spins must be up for the current to flow.

For this example, the display saying " = 2" will be illuminated because current from #2 flows through the channel made by

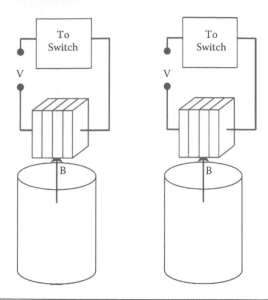

Figure 7.18 More current (or voltage) passes through the layered structures when the magnetic fields are up.

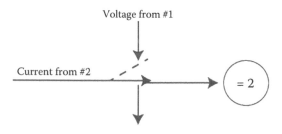

Figure 7.19 The combined signals from the two memories allow the computer to deduce that the sum is 2.

conducting the signal from #1. The toy computer has successfully determined that 1 + 1 = 2.

The inadequacy of this computer is obvious. Comparisons with a real computer illustrate the amazing technology of modern computation:

1. The toy computer memory consisted of only 2 bits. The memory of a real computer is typically trillions of bits. All these magnetic bits are packed onto a hard disk less than 10 centimeters across. The magnets that store the 1s and 0s have dimensions on the order of only hundreds of atoms (~10 nanometers). The hard disk rotates rapidly.

2. The toy computer wrapped a wire around each bit to produce the magnetic polarization. In a real computer, the numbers (0 and 1) are stored sequentially. As each bit on the hard disk passes a magnetizing "write head," the appropriate signal is applied. A bit is magnetized as it flies past a write head. This requires extraordinarily accurate high-speed timing and positioning.

3. Two circuits were used to extract the numbers on the toy computer. In a real computer, numbers are extracted in sequence from the bits on the rotating disk using a single circuit that utilizes giant magnetoresistance.

4. The numbers were sent directly to the toy CPU. Because numbers are obtained sequentially in a real computer, they are generally stored in a different place as charges (+ or –) until they are needed.

5. The toy computer was required to combine only two numbers to obtain a result. Real computers require a great deal of extra logic, not only to add bigger numbers, but also to "direct traffic." The computer needs to know which bit to read, what operation to perform, and what to do with the result. Considerable programming and bookkeeping are needed. These traffic instructions are also stored, mostly on the hard disk. A computer essentially teaches itself what to do each time it is turned on. This makes the lethargic startup of a computer less surprising, if not less annoying.

6. The CPU of the toy computer consisted of a single transistor. The CPU of a real computer contains millions of transistors, and they are all packed into a small area only a few centimeters across. The basic operations, like adding 1 to 1, are done sequentially and rapidly, typically more than 100,000 basic operations are done each second.

Despite these differences, the toy computer has many of the basics of all computers. It stores data, transfers data, and performs logical operations on the data.

Magnetism is not the only choice for keeping track of yes-no questions. Choice depends on whether the data are to be stored for long or short times, how much space is available, how quickly data are accessed, and requirements for ruggedness and reliability. Various alternatives to magnetic memories appear in the computers embedded in many smaller electronic devices.

7.3.6 Sound Reproduction

7.3.6.1 Edison's Phonograph When Thomas Edison first shouted, "Mary had a little lamb," into his new device, he was pleasantly surprised when it reproduced his words (1877). His remarkable achievement was surely not as fundamental as the invention of writing. Still, there is an analogy.

Writing was invented more than 5000 years ago. It encodes a sequence of words (in time) into a string of symbols (in space). The space-to-time transformation occurs because we move our hand (often from left to right) as we transcribe language. Our familiarity with the written word means we fail to notice the level of abstract reasoning needed to accomplish this time-to-space transformation. Reading is the inverse process that transforms symbols in space back into a language expressed in time.

Edison's phonograph (which resembled that in Figure 7.20) took the time-to-space transformations from the process of word transcription to the much shorter time intervals associated with the pressure oscillations of audible sound. Consequently, much more information

Figure 7.20 An early phonograph that employed physics similar to Edison's original design.

is captured, including the subtler qualities of speech and music that cannot be reproduced by even the most sophisticated writer.

The key to recording sound is encoding the sound wave, typically oscillating at hundreds of cycles per second (hertz). This is accomplished by using the sound wave pressure to move a physical object. The most important object moved by sound is the eardrum, and part of Edison's invention resembled a giant ear. The outer part of the ear became a tapered tube that concentrated the sound. The eardrum was replaced by a thin flexible diaphragm. A pin attached to the diaphragm was analogous to the outermost ear bone. This pin moved back and forth in synchrony with the oscillating pressure.

The rest of Edison's phonograph was not ear-like. The moving pin pushed against a surface (foil or wax) that could be permanently indented. The sound vibrations were separated in space because the surface moved past the pin as the recording took place. The moving surface was initially a rotating cylinder attached to a screw, so the sound was transformed into a wiggly line of indentations etched out on a helical path. The record was permanent even though the initial sound vanished as soon as Thomas finished saying "Mary had a little lamb."

At any later time, the sound could be reproduced by inverting the physics. The motion of a second needle that followed the track of the indentations on the cylinder moved a diaphragm, which in turn moved the air and produced sound. The reproduction could be saved for a long time.

Phonographs and their generalizations have been greatly refined since the time of Edison. Electronics greatly improved sound recording. The improvements are based on the conversion of physical motion into an electric signal (e.g., microphone) and the reverse process of converting the electricity into mechanical motion (e.g., speaker). Electrification did not take long. The telephone changes sound to electric signals, and it was patented by Alexander Graham Bell a year before Edison's announced discovery of the phonograph. As is usually the case, the history of these inventions is really more complicated than the names of two famous American inventors would suggest. There is even more controversy about who invented the telephone than who invented the phonograph.

There are many ways that mechanical and electrical signals can be interchanged. A simple example, but not the most practical or

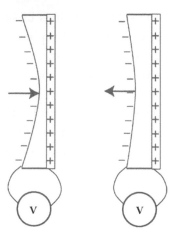

Figure 7.21 The position of the flexible diaphragm and the voltage *V* vary together.

common example, is illustrated in Figure 7.21. Two views of a pair of charged surfaces are shown. One of the surfaces is flexible. Equal and opposite charges on the surfaces bends the flexible surface. The surface can also be bent by the pressure of a sound wave. The surfaces are connected by wires to *V*. A sound wave can change the voltage (microphone case), or a changing voltage can produce a sound wave (speaker case).

Consider first the case in which Figure 7.21 describes a microphone. The pressure of a sound wave moves the flexible surface back and forth at the sound frequency. If the charges on the two surfaces are fixed, the voltage will vary because it is proportional to the average separation between the surfaces. (Fixed charge means the electric field is constant, and the voltage is the product of the field and the distance.) In this case, variations of the voltage *V* are produced by the sound wave.

Alternatively, the system in Figure 7.21 describes a simplified electrostatic speaker. The *V* is now a source of a varying voltage. A large voltage produces more charge and a larger attraction between the surfaces, hence causing greater bending. The back-and-forth motion of the flexible surface follows the voltage and produces the sound wave.

There is an enormous advantage in changing the physical vibrations of sound waves into electrical signals. The signals can be amplified by circuits that are sophistical analogues of Figure 7.16. Very small oscillations on a vinyl record can be changed into very loud sounds.

Despite the improvements of sound reproduction resulting from the electrification and amplification of audio signals, the process is far from perfect. Two serious problems are associated with recording sound on a vinyl disk or magnetic tape.

Problem 1: Sound frequencies are typically hundreds of hertz, so even 1 minute of music results in many pressure oscillations. Recording these is reasonable only if they are compressed into a relatively small space. The curve representing sound is thus a fine and delicate line.

Problem 2: The wave amplitude of a loud sound has at least 10,000 times the amplitude of a faint whisper. It is hard to capture these extremes.

Sound's rapid oscillations and extreme intensity variations mean reproduction that is compact, durable, and faithful at all audio frequencies and intensities is a real challenge. Digital recording goes a long way toward meeting this challenge.

7.3.6.2 Digital Recording Digital represents a fundamental change in the way sound (and much more) is preserved. At first glance, replacing a curve representing a sound wave by a series of numbers seems cumbersome and inherently inaccurate. The numbers represent the amplitude only at specific times. No information about the sound is preserved in the intervals between the selected measurement times. The end run around this problem is achieved through speed. The sound amplitude is sampled and recorded so frequently that virtually no useful information is lost. The expansion of the timescale is suggested by Figure 7.22. Typically, the sound amplitude is obtained 44,100 times a second. This rapid sampling rate means frequencies up

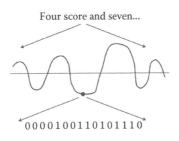

Four score and seven...

0000100110101110

Figure 7.22 Changing language to code. Writing, analogue recording, digital recording.

to about 20,000 hertz can be accurately reproduced. Because people cannot hear higher frequencies, this rate is sufficient. If one wished to record music that could entertain bats, dolphins, or even the family dog, 44,100 samples per second would be inadequate.

To obtain a digital signal, a microphone changes the sound pressure to a voltage. The voltage is changed into a number by an electronic analogy of the game of 20 questions. For example, if the voltage is between 0 and 7 volts, one assigns an integer to the voltage by finding the answer to three questions.

1. Is the voltage 4, 5, 6, or 7? If the answer is yes, write the number 1. A no answer receives a 0.
2. Is the voltage 2, 3, 6, or 7? Write the 1 or 0 for this question after the first answer.
3. Is the voltage 1, 3, 5, or 7? Use the answer to write the third digit.

The answers to the three questions determine the voltage as follows:

0: 000
1: 001
2: 010
3: 011
4: 100
5: 101
6: 110
7: 111

This correspondence between numbers and a sequence of 1s and 0s is called *binary* numbering, and each binary digit is called a *bit*. Electronics circuits can quickly ask the yes-no questions posed, so a voltage can be changed to a binary number in a remarkably short time.

The speed of digital recording is a feat only achievable because of modern electronics. At each sampling time, the electronic answer to 16 yes-no questions determines the voltage to be one of 65,536 (2 multiplied by itself 16 times) values. Answering 16 yes-no questions 44,100 times a second means (neglecting some bookkeeping) that the system is reading 705,600 bits each second. Double this rate is needed when recording two channels of sound (stereo).

7.3.6.3 *Digital Playback*   One might think that a digital recording could be too much information to store anywhere. What could one do with the 5 billion numbers needed to digitize a 1-hour stereo recording? Thanks to microelectronics, 5 billion is no longer a big number. It is remarkable that all these bits can be encoded and played back on a single compact disc just a few centimeters across.

Size is the key to packing a huge amount of data on a compact disc. Lasers are used to read the numbers. In principle, that means the spacing between bits on the disk's surface need not be not much greater that the wavelength of the light used to read the disk. Because the wavelengths of the laser lights used are less than one-millionth of a meter, one could in principle store 100 million bits on just 1 square centimeter of surface. Thus, compact discs are not magic, even though they do rely on technology that seems magical.

The bits are read from the disk at the same rate they were stored, 1,411,200 bits/second. This is speed reading to the extreme. The disk rotates under a narrowly focused laser at the recording rate. Each bit either is reflecting (1) or is not (0). The reflection is not determined by making light and dark spots on the disk. Instead, each bit has a left and right side. If the two sides are at the same height, their reflected light is added. If one side is depressed so it is a quarter wavelength lower than the other side, the lights reflected from the two sides cancel each other, and a reflection is not observed. This is another generalization of interference of the type seen in Young's double-slit experiment (Figure 5.24 of Chapter 5).

For most listeners, some data can be lost and the sound will still seem satisfactory. A commonly used method of compacting the data by about a factor of 10 is called MP3. At the other extreme, there are a few people who strive for near perfection, and they spend a great deal of extra money to obtain recordings with significantly higher data density.

Changing the digital data back into audible sound remains the weak link in the chain of sound reproduction. Electrostatic speakers are essentially microphones working in reverse. Changing voltage between the two plates makes them oscillate. Refinements and scaling make this a successful, but expensive, technique. More often, the electronic signal is changed to a current that powers a magnet that pushes on a permanent magnet, causing motion. Resonance is one of the problems associated with sound reproduction, and engineering tricks must be used to prevent speakers from being too loud at their resonant frequencies.

We lose our ability to hear high frequencies as we age. For many of us, by the time we can afford the highest-quality sound reproduction, we won't be able to notice the difference.

8

RELATIVITY

8.1 Introduction

Special relativity came to Albert Einstein fairly quickly. It was a "stroke of genius."

General relativity was entirely different. Einstein labored intensely for many years to produce his final elegant picture of the universe. He made mistakes along his path to one of the most remarkable scientific leaps of all time. General relativity is an extension of special relativity that incorporates gravity. It remains the foundation of our understanding of the large-scale universe.

Special relativity is based on a fundamental observation.

Motion at a constant speed does not change physics. In particular, a measurement of the speed of light is the same no matter how fast the observer is moving.

General relativity is based on a second fundamental observation called the *equivalence principle.*

Acceleration is essentially the same as gravity.

8.2 Special Relativity

8.2.1 A Famous Equation: Mass and Energy

The book *Einstein's Mistakes* found flaws in all seven of Einstein's proofs of

$$E = mc^2$$

Why did he so carelessly present the only physics formula that "everybody" knows? This is an example of physical insight transcending

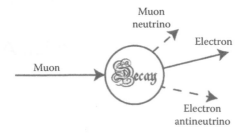

Figure 8.1   The mass of the muon is sufficient to allow its decay into an electron and two neutrinos with a great deal of energy to spare.

technicalities. Criticism of Einstein's mathematics shows the difficulty in rigorously proving revolutionary ideas.

The equation $E = mc^2$ says a small mass m can be changed into a great deal of energy E. The ratio of energy to mass is the square of the speed of light c. The decays of the muon and the neutron are illustrative examples.

The unstable muon becomes an electron (and two almost invisible neutrinos) with a half-life of 2.2 microseconds, as pictured in Figure 8.1. The muon has 200 times the mass of an electron, and neutrinos are nearly massless. Applying $E = mc^2$ means muon decay yields more than 100 million electron volts of kinetic energy that is shared by the electron and the neutrinos. This decay energy dwarfs chemical energies, which are typically only a few electron volts. Muon decay and nuclear beta decay are both examples of weak interaction physics, which are described in Chapters 9 and 10.

8.2.1.1 Beta Decay   Beta decay is a form of radioactivity seen when a nucleus ejects an electron instead of an alpha particle or a photon (gamma ray). Beta decay is mysterious because the nucleus contains no electrons.

The clearest example of beta decay is seen when an isolated neutron changes into a proton, an electron, and an antineutrino. This is possible because the neutron mass is slightly larger than the sum of the electron, proton, and neutrino masses.

neutron → proton + electron + antineutrino

This decay has a half-life of about 10 minutes for an isolated neutron. Using $E = mc^2$ tells us that 0.78 million electron volts of kinetic energy are released in the decay.

Because neutrons decay, one might wonder why there are any left. The stability of neutrons in nuclei is a result of the strong nuclear force that binds the protons and neutrons. The binding energy is so large that neutron decay within a nucleus would increase, rather than decrease, the total mass, making beta decay impossible. Most, but not all, nuclei in our environment are stable. If undisturbed, they will live forever (probably).

8.2.1.2 Neutrinos   No one knew of the neutrino when nuclear beta decay was first observed. Thus, when the total kinetic energy of the decay particles was measured, energy seemed to have disappeared. The energy was missing because no one could detect the neutrino, let alone measure its energy. Quantum mechanics is so strange that some (notably Niels Bohr) believed energy conservation might not strictly apply for the smallest distance scales. Wolfgang Pauli was convinced that energy conservation could never be violated, even for the submicroscopic atomic nuclei. He salvaged energy conservation by postulating an invisible particle. The particle was later named the *neutrino* by Enrico Fermi, who also refined the theory. It is not surprising that some were skeptical of Pauli's particle and its unseen energy. Fortunately for physics, the neutrino is not quite invisible. Twenty-six years after Pauli's 1930 suggestion, an experiment detected the elusive neutrino. The experiment was far from easy, and its discovery earned a Nobel Prize for one of the experimenters. Pauli also received a Nobel Prize, but it was for the Pauli exclusion principle. Today, neutrino measurements have revealed new physics, but the experiments remain challenging because neutrinos usually pass through everything without leaving the faintest footprint.

8.2.2 The Energy Triangle: Nothing Goes Faster than Light

The energy triangle is a generalization of the famous $E = mc^2$ for moving particles.

The total energy of a free particle includes the mc^2 "rest mass energy" as well as the kinetic energy resulting from the particle's momentum. The triangles of Figure 8.2 show the relativistic relation between momentum (vertical side), mass (horizontal side), and energy (hypotenuse). For the nonrelativistic case on the left, the small momentum

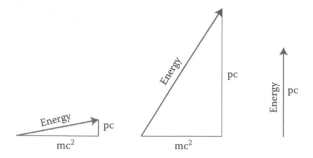

Figure 8.2 The relation between mass, momentum, and energy. The ratio of the particle's speed to the speed of light is equal to the ratio of the *pc* side of the triangle to the *Energy* side of the triangle.

means the kinetic energy is the nonrelativistic $(1/2)mv^2$. The center triangle shows the relativistic case in which large momentum produces a total energy only slightly larger than the product of the momentum and the speed of light, or *pc*. The line on the right of Figure 8.2 is a triangle whose base has disappeared. It describes the relation between energy and momentum for a particle with zero mass. The photon has no mass, so its energy is exactly *pc*.

Pythagoras is the name associated with the right triangle rule. Applying his formula to the energy triangles of Figure 8.2 gives the equation

$$(Energy)^2 = (mc^2)^2 + (pc)^2$$

The energy triangle also determines the particle speed. The ratio of this speed to the speed of light v/c is the same as the *pc* side divided by the *Energy* side of the triangle (*pc/Energy*). For the nonrelativistic triangle on the left, that means the speed is essentially the momentum divided by the mass. For the center case of large momentum, the speed is close to the speed of light. However, no matter how large the momentum becomes, the speed is always a little less than *c*. The exception is the zero-mass photon on the right. The collapsed triangle means the photon speed is always equal to the speed of light (in a vacuum).

One must be a little cautious about the rule that "nothing goes faster than light." General relativity allows space itself to change in time. Speed comparisons make sense only when objects are described by the same coordinates. This means the light speed limit applies for objects passing each other, but one must avoid extending this speed limit to objects in distant parts of the universe. This is an important distinction when trying to understand the "inflation" modification of the "big bang" theory. Inflation can appear to violate the absolute speed limit.

8.2.3 Speed Is Confusing

Two rules of relativity lead to real conceptual difficulties: (1) Nothing moves faster than light; and (2) the speed of light is absolute. Figure 8.3 illustrates a common-sense idea that appears to violate these rules.

In principle, it is possible to build a rocket ship that could move away from Earth at half the speed of light. In principle, people on that rocket could build a smaller ship that would move away from the mother ship in the same direction, again at half the speed of light. Common sense says the daughter ship would be moving at the speed of light c, which violates a basic precept of relativity. Even worse, it is also possible that the daughter ship could build a still smaller ship moving away again at half the speed of light.

This apparent superluminal contradiction appears because speeds cannot be simply added. The problem can be traced back to the definition.

$$\text{speed} = \frac{\text{distance}}{\text{time}}$$

Speeds cannot be added because relativity says strange things happen to both time and distance in moving systems.

8.2.4 Time and Simultaneity

Relativity asks us to reconsider the meaning of time. Time is no longer absolute, and "simultaneous" is no longer a valid concept. Special relativity tells us that two people moving past each other can honestly disagree about which of two events happened first.

As a contrived example of time ambiguity, imagine that the sun explodes at noon (an unlikely event). Just 250 seconds later, a rocket

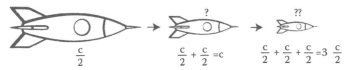

Figure 8.3 A problem with the addition of velocities.

leaves Earth for a distant planet. The people who escaped the explosion were lucky. No earthling could know about the solar catastrophe until 500 seconds past noon because it takes light 500 seconds to reach us from the sun, and no signal travels faster than light.

Relativity says that the 250-second time difference between the solar explosion and the rocket departure is not absolute. Another preposterous assumption can illustrate this.

Assume aliens monitoring our solar system are moving from sun toward Earth at half the speed of light when the sun explodes. The sun, alien craft, and Earth are shown in Figure 8.4. Because of their motion, the aliens think the two events (solar explosion and rocket departure) happened at exactly the *same* time. The aliens did not make a timing mistake. The different time intervals are one of the curiosities of relativity. If the aliens had been moving even faster, it would appear to them that the space ship left *before* the sun exploded. However, no matter how fast they traveled in either direction, they would always know that neither event occurred early enough to cause the other. Relativity assaults our intuition, but it does not defy logic. No signal from the sun could reach Earth in time to warn the rocket. No signal from the rocket launch could tell the sun to explode.

One can extend the silly example even further and consider the case for which causality is possible. Assume that the spaceship left Earth 1000 seconds before the sun exploded, and the ship's inhabitants shot an exploding magic bullet to the sun just at liftoff. Because light can travel from Earth to the sun in only 500 seconds, a relativistic magic bullet could initiate the solar explosion. Furthermore, the aliens observing the two events (spaceship departure and sun explosion) would always observe that the spaceship departure happened early enough to cause the solar explosion. That is, if causality is possible in one coordinate

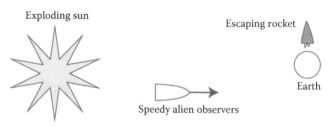

Exploding sun

Escaping rocket

Earth

Speedy alien observers

Figure 8.4 Which happens first, the exploding sun or the escape from Earth? It depends on the speed of the observer.

system, it is possible in all coordinate systems. Causality remains sacred even though time takes a beating in special relativity.

Time ambiguity can be confusing, but it can never be extended to the ridiculous. No moving coordinates will allow Abraham Lincoln to exchange jokes with George Washington. Time travel, violation of causality, and other mind-bending impossibilities are not consequences of special relativity.

Relativity tells us that motion distorts space as well as time. The example of muon decay that follows shows that distorted time in one coordinate system manifests as distorted space in another coordinate system.

8.2.5 Muons: Time Dilation and Length Contraction

A muon is essentially a heavy, but unstable, electron. Given a large number of muons, half will change into electrons (plus neutrinos; see Figure 8.1) within only 2.2 microseconds. Even though short-lived muons are produced high in the sky, we are not free of these particles. One muon passes through every person's head about once a second. Relativity explains why. Relativistic time dilation and length contraction are the responsible mechanisms.

Cosmic rays produce the muons 10 kilometers high, near the top of our atmosphere. These muons move at nearly the speed of light. Multiplying the light speed $c = 399{,}792{,}458$ *meters/second* by the 2.2-microsecond muon half-life gives an average travel distance of only 660 meters. This means the muons should not reach Earth's surface—but they do.

The surprising shower of muons on Earth is a consequence of relativity's distortion of time. A person on Earth observes that the muon's lifetime is increased by the ratio *Energy/* (mc^2). If the muon's total energy is 20 times its rest mass energy, it can travel (on average) 20×660 meters = 13.2 kilometers instead of 660 meters. The increased lifetime allows many muons to reach the ground. We are bombarded by muons because of this relativistic effect, called *time dilation*.

Relativity alters measurements of space as well as time. This can be seen in the same muon example. From the muon's point of view, its life expectancy is an unchanged 2.2 microseconds, but the muon perceives the distance to Earth to be shortened by the same relativistic ratio *Energy/* (mc^2). This apparent shrinking of travel distance is called *length contraction*. Time dilation and length contraction are equivalent ways of explaining why so many muons traverse our bodies every day. Both explanations are correct. Whether time is dilated or length is contracted depends on your coordinate system.

Speed must be close to the speed of light before time dilation or length contraction becomes significant. For example, the fastest human speed relative to Earth was accomplished by the astronauts on Apollo 10. Their speed of 11.1 kilometers per second (more than 24,000 miles per hour) is impressive, but it is only 27 millionths of the speed of light. At this speed, the length of an astronaut's arm would appear to shrink by only the diameter of a single atom. Even moving at 10% of light speed produces only a 0.5% contraction.

Jesus said:

> Again I say to you, it is easier for a camel to go through the eye of a needle than for a rich man to enter the kingdom of God.

Rich people might consider length contraction as a promising alternative to renouncing excessive wealth. The right-hand camel shown in Figure 8.5 is traveling at 87% of light speed. It appears shortened by a factor of 2. Perhaps additional speed could help the camel pass through the eye of a needle. Alas for we rich folk, this will not work. The camel contraction occurs only along its direction of motion. The camel's height is unchanged, making needle penetration impossible.

Figure 8.5 The apparent shrinkage of a camel moving at 87% of the speed of light is only along the direction of motion. The camel's height is unchanged, and it cannot be passed through the eye of a needle at any speed.

A confusing aspect of camel length becomes apparent when the situation is considered from the moving camel's point of view. The camel on the right of Figure 8.5 thinks the camel on the left is shortened. The apparent paradox is resolved by considering the method of measurement. When a camel is speeding by, its length is determined by making a *simultaneous* observation of the coordinates of the camel's head and tail. But, simultaneous is ambiguous in relativity. The camel thinks you find him shortened because the position of the head was measured *before* the position of the tail, and in that time interval, the tail moved on.

8.2.5.1 Space Travel   "Warp speed" (or ludicrous speed) is impossible. Nothing, not even a spaceship supplied with a fictitious antimatter drive, can travel faster than light. However, time dilation can be a big help in a one-sided way. In principle, an astronaut could travel to the nearest Alpha Centauri star system, which is four and one-third light years away, while aging only one month. To do this, the astronaut's speed must be 99.875% of the speed of light, so time dilation slows the astronaut's clocks. If the astronaut returns home traveling at this same outrageous relativistic speed, he or she will only be two months older. However, people on Earth do not experience the advantage of time dilation. They will rightly observe that the space traveler did not exceed the speed of light, so they will think the trip took eight and two-thirds years. Everyone one Earth will have aged eight and two-thirds years even though the astronaut aged only two months. This curiosity in aging is called the *twin paradox* even though it is not really paradoxical.

Not even the most starry-eyed scientists believe that people can achieve 99.875% of the speed of light. We must resign ourselves to life within our solar system. If aliens live on different stars, it would take lifetimes to pay them a visit.

8.2.6 Relativity Beyond Mechanics

Special relativity is a theory of space and time that abandons simultaneity but maintains the constancy of the speed of light. There is more. The same type of rules that manipulate space, time, energy, and momentum applies to other physical quantities. There are analogous relations for electromagnetism that relate charge density and electric current density. Slightly more complicated formulas tell how electric and magnetic fields are altered in moving coordinate systems. The relativistic form of Maxwell's equation yields an invariant speed of light. The laws of electromagnetism remain valid and unchanged in all moving coordinate systems. Quantum electrodynamics is a successful relativistic version of quantum mechanics that incorporates electromagnetism. It is briefly described in Chapter 10.

8.3 General Relativity

Special relativity does not adequately describe gravity. This is the job for general relativity.

The story of general relativity starts with Einstein's groundbreaking theory of 1916. The central equations of general relativity determine the structure of space and time in terms of the energy and matter residing in space. Relativity pioneer John Wheeler expressed the essence of general relativity with the paraphrased statement: "Space-time tells matter how to move and matter tells space-time how to curve."

Before considering the cosmological implications of general relativity, two examples (the gravitational red shift and light bending) suggest the unusual ideas. An honest presentation of general relativity is both difficult and subtle. The following examples are not subtle, and they can be misleading if taken too literally.

8.3.1 Red Shifts and Black Holes

8.3.1.1 Red Shift General relativity tells us that the frequency of light decreases as it moves away from a massive object. This "gravitational red shift," suggested by Figure 8.6, differs from the Doppler red shift that occurs when the source and observer are separating.

If a ball is thrown up, gravity pulls it down, and its kinetic energy decreases. General relativity says a similar energy loss applies to light. One could argue that gravity should have no effect on light because photons are massless. However, photons do have energy, and Einstein's famous $E = mc^2$ notes the equivalence of mass and energy. If one assumes that the m in Newton's gravity formula $F = GmM/r^2$ should be replaced by its energy equivalent E/c^2, then it makes sense that

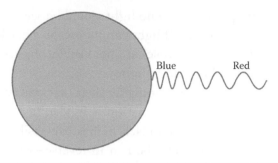

Figure 8.6 A simplified and exaggerated gravitational red shift. The big sphere is a massive object, and the curve represents a light ray being shifted to longer wavelength and lower frequency.

photons should also be subjected to gravity. As the photon loses energy pulling away from the mass, its energy and frequency f will decrease because the energy of a photon is hf, where h is Planck's constant.

The famous Pound–Rebka experiment of 1959 confirmed the gravitational red shift to about 10% accuracy. Light was beamed up 22.6 meters, and its frequency at the two heights was compared. It is a credit to the experimenters (and the Mossbauer effect described in Chapter 9, Section 9.3.5) that the gravitation frequency change of the light (about two parts in 10^{15}) could be measured.

A change in the frequency of a light makes one think again about time. If a light with frequency f is turned on for a time T, the total number of oscillation is the product $f \cdot T$. The observer at the height d surely sees the same number of oscillations. Because the product $f \cdot T$ is fixed, a smaller f means a larger T. General relativity is telling us that clocks up high are faster, and this elongates the time interval T. The general conclusion is that time depends on both relative motion (special relativity) and the proximity of massive objects (general relativity).

Global positioning systems (GPSs) rely on precise timing to locate objects to within meters. The gravitational red shift means the clocks on the orbiting GPS satellites are fast by about 45 microseconds per day, or about a 60th of a second in 1 year. In principle, general relativity corrections could introduce significant errors in the position locations, but consistency checks mean GPSs would work even if the world was ignorant of general relativity.

8.3.1.2 Black Holes The gravitational red shift suggests the possibility of a black hole. If an object were so massive that a photon lost all its energy (and frequency), then no light would be seen coming from this compact massive object. If light cannot escape a black hole, then nothing can. Although black holes are not visible, objects falling into a black hole produce abundant radiation.

Black holes are characterized only by their mass, angular momentum, and charge. No other information is hidden in the black hole. One of the many mysteries of a black hole is irreversibility. Without black holes, physical laws could be run in reverse—at least in principle. Reversibility implies information is never lost—it just becomes scrambled. Black holes muddy the water because one cannot run time backward for objects captured by a black hole. Another black hole complication relates to quantum tunneling and the electron-positron creation that could eventually cause the evaporation of a black hole. Black hole evaporation was first considered by Stephen Hawking.

Black hole radiation presents us with an "information paradox." Consider the case where two different galaxies are completely absorbed and become two black holes. Even if the galaxies looked very different, the two black holes would be identical (assuming only the same mass, charge and angular momentum). Then if one waited long enough for the black holes to evaporate, the results would be indistinguishable. All information that characterized the differences between the galaxies would be lost. Losing information like this does not sit well with physics. Some believe that string theory will resolve this paradox. Others are not so sure.

8.3.2 Equivalence Principle and Light Bending

The equivalence principle was Einstein's foundation for general relativity. It says a system freely falling in gravity "feels" no gravity. An experiment done in a box accelerating because of gravity must yield the result expected at rest in a gravity-free environment. The falling observer in the left of Figure 8.7 is doing such an experiment. The observer sees an unbent light beam. The stationary observer outside the falling box, shown on the right of Figure 8.7, must see the light arrive at the same spot on the box, but this appears to be a bending path for the light, as if gravity pulled on light. Light travels so fast that the bending is usually insignificant. Earth's gravity does not send our car headlight beams into the ground.

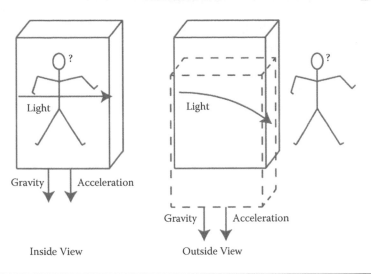

Inside View   Outside View

Figure 8.7   The equivalence principle suggests the bending of light, but correct results need more than this simple model.

The falling object analogy has some problems that lead to ambigui-ties and an underestimate by a factor of 2. Einstein's numerically correct prediction was verified by an experiment showing that the sun bends starlight. Normally, a solar deflection of starlight is invisible because of the blinding sun, but the solar eclipse of 1919 made the experiment possible that validated Einstein's theory and enhanced his fame.

The bending of light by massive objects has progressed from a the-ory, to a discovery, to an experimental technique. Light from a distant source that passes by a massive system like a galaxy cluster will bend. In Figure 8.8, an observer sees the light from a distant star in two directions because of the light bending. The star may appear as two images, a ring, or some other distorted image. In other cases, one may

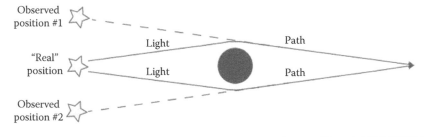

Figure 8.8   Light bending around a massive system can have a double image as in this two-dimensional representation. In three dimensions, the image of the distant star may appear as a ring or a more complicated shape.

see only a change in the light intensity. This "gravitational lensing" has become a tool of astronomy.

8.3.2.1 Curved Space A carpenter sights along a board to see if it is warped. Carpenters never consider the possibility that the light beam rather than the board is warped. A light path (in vacuum) is essentially the definition of a straight line. One can interpret light bending by massive objects as an indication that space, rather than light, is bent. Indeed, space and time do not have a simple geometric structure in general relativity.

An example suggesting curved space is illustrated by the rapidly rotating disk of Figure 8.9. According to special relativity, lengths of moving objects are shortened. The camel example (Figure 8.5) shows contraction only in the direction of motion. This means the circumference of a camel carousel should be shortened while the radius remains unchanged. The circumference will no longer be 2π times the radius. Of course, this is impossible in the world that we know. But, if space is curved, all bets are off for the relation between radius and circumference.

8.3.3 Cosmology

8.3.3.1 Big Bang Basics The more distant a galaxy is, the faster it is moving away from us. The Doppler shift from blue to red determines the speed of recession. The distance is harder to determine. A great deal of the history of astronomy is related to distant measurements. Knowing the distance and the speed of recession of two galaxies allows one to determine the time when they were at the

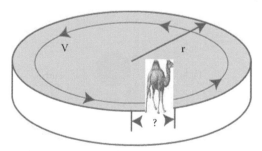

Figure 8.9 Contraction in the direction of motion of the rotating disk suggests that the circumference is not 2π times the radius.

Figure 8.10   None of these is a realistic image of the big bang, which really cannot be visualized.

same place. Essentially, all galaxies move away from each other with speeds proportional to their separation (with some minor corrections). This means there was an earlier time when *all* the objects were at the same place. This is evidence for the "big bang" view of the universe. Comparing the speeds determined by the Doppler effect with the observed distances means the big bang occurred 13.8 billion years ago (roughly). None of the cartoons in Figure 8.10 captures the big bang because it cannot really be drawn. This event created all the matter in the universe. It also created the space and time in which the matter resides. No one knows what happened before the big bang (if anything), but that does not keep some people from speculating.

8.3.3.2 Mathematics of the Big Bang   Einstein's general relativity provides the mathematical framework of the big bang. As one might expect, equations describing the mutual interaction of curved spacetime and matter are too difficult to solve for a general case. It makes sense to simplify. A reasonable assumption ignores the stars and galaxies and replaces them by an average density. The smoothed universe is assumed to be homogeneous and isotropic. Versions of this "cosmological principle" were stated by Newton, Einstein, and others. The cosmological principle appears to be justified by observations of approximate uniformity on the largest distance scales, but the evidence is not conclusive.

Solutions of Einstein's equations based on the cosmological principle were independently developed in the 1920s by Monseigneur Georges Lemaitre, Alexander Friedmann, Howard Robertson, and Arthur Walker. Some of the conclusions are as follows:

1. The universe is expanding (or contracting).
2. For an expanding universe, there is an initial time ($t = 0$) that corresponds to the beginning of the expansion. This is the big bang. A contracting universe would eventually undergo an analogous "big crunch."
3. Expansion means the light from distant objects moving away from us will be Doppler shifted to lower frequencies. (Blue becomes red.)
4. Light takes time to travel, so we see distant objects at earlier times.
5. No light was produced before the big bang, so we cannot see objects so distant that light would take more than 13.8 billion years to get here.

Initially, Einstein did not find a big bang or a big crunch appealing. When he learned of Alexander Friedmann's work describing an expanding universe, he was not favorably impressed—at least not at first.

8.3.3.3 Cosmological Constant   Even before the formal mathematics of the big bang was developed, Einstein, Newton, and others realized that gravity could collapse the universe. This was a problem in 1917 because the universe was viewed as static and unchanging. By adding a "cosmological constant" to his equations, Einstein managed to balance attraction with repulsion and produce a static (but not stable) model of the universe that avoided the big bang.

Edwin Hubble published results in 1929 that showed larger red shifts for more distant objects. Hubble was not the first to observe red shifts of distant galaxies, and strangely, he had trouble accepting an expanding universe. But, the evidence for expansion was clear. Suddenly, the big bang was not such a silly idea. The cosmological constant was no longer necessary.

Anyone who has never made a mistake has never tried anything new.

This statement is widely attributed to Albert Einstein. If he did not say it, he should have. It was after the discovery of an expanding universe that Einstein called the cosmological constant his "biggest blunder." (This blunder statement may have been manufactured, but it sounds good.) If Einstein had insisted on his simpler, bare bones version of general relativity, he could have predicted an expanding (or contracting) universe.

The cosmological constant was never really dumped into the trash bin of mistaken ideas. Its most recent resurrection is associated with the ambiguously named *dark energy*. Recent observations showed that the universe expansion is accelerating, and the acceleration is blamed on dark energy. This nonintuitive result is confusing because one would think that gravity should eventually pull everything back together. Instead, there appears to be an extra mysterious "something" that pushes everything apart. The cosmological constant can again be called on to fit the observed acceleration. As with Einstein's first introduction of the cosmological constant, picking a number to fit the data is not satisfying. A surprisingly large number of people are busy devising models that give the cosmological constant a more physical interpretation. If anyone succeeds, it will be big news.

8.3.3.4 Cosmic Microwave Background Electromagnetic radiation permeates the universe. Every planet, star, and galaxy is bathed in microwaves. The radiation extends over a wide frequency range. A graph of the radiation intensity as a function of frequency has the shape of the thermal radiation curve described in Chapter 6, Figure 6.4, associated with a temperature of 2.725 K. This is very cold, but the typical microwave frequency is more than 50 times the frequency of a cell phone or a microwave oven. This radiation was discovered, partly by accident, in 1964. The microwaves are considered the strongest experimental evidence for the big bang theory.

8.3.3.5 Motion Through the Universe We can identify Earth's velocity through the cosmic radiation by noticing that the average frequency and intensity of this microwave background is increased in the direction we move, as suggested in Figure 8.11. Earth's velocity relative to the microwaves is 600 kilometers/second toward the constellation Leo. This seems like a pretty high speed, but it is only 1/500 the speed

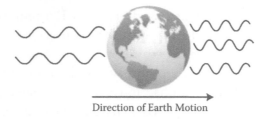

Direction of Earth Motion

Figure 8.11   The motion of Earth with respect to the background microwave radiation shifts its frequency and intensity.

of light. To a pretty good approximation, we are at rest with respect to a radiation that permeates the universe.

Earth is not special. Mysterious beings living on a distant galaxy receding from us at half of light speed would also find themselves to be nearly at rest with respect to the microwaves. Teenagers living on any galaxy can justifiably think of themselves as being the center of the universe. How can two objects moving rapidly away from each other see essentially the same microwave environment? The answer lies in the peculiar nature of the cosmic expansion. It is not as simple as everything running away from a central point. The cosmological principle says every point is equivalent, so every point appears to be a central point that is at rest with respect to the background radiation.

8.3.3.6 Origin of the Radiation   The microwave background is believed to be the remnant of the high-frequency radiation of a very hot universe that was produced shortly after the big bang. Because everything is moving apart, this high-frequency radiation has undergone an enormous Doppler shift that reduced the radiation frequency to microwaves. One can understand this frequency drop (roughly) with a simple model. Assume the universe is made of small boxes with reflecting walls. Because each box is identical and none is in a preferred position, the reflections do not change the character of the radiation in the boxes. Universe expansion means each box is expanding. Because the walls of the box are receding from each other, light's frequency is lowered by the Doppler shift each time it is reflected from a wall. The huge expansion that has taken place since the photons were first released explains the low frequency.

8.3.3.7 The Horizon Problem The universe is too big to see. Two examples illustrate this:

1. Even with the best of telescopes, we can only see a finite distance. Light takes time to travel, so looking into the distance is also looking into the past. There is no past beyond 13.8 billion years (the time of the big bang), so we cannot see a part of the universe whose light would take more than 13.8 billion years to reach us. This is our "horizon." One might guess that this horizon corresponds to the end of the universe, but this does not follow. By simply waiting another billion years, a very aged astronomer would be able to see further into the past and a bigger universe.

2. Cosmic microwave radiation arriving at Earth from point N in the direction of the North Star was produced by light from the most distant source we can detect. This source is at (or nearly at) our horizon. Cosmic microwaves coming from point S in the opposite direction were also produced by objects near our horizon. The two horizon points are separated by twice the horizon distance. There is no way light could travel twice the horizon distance in 13.8 billion years, so light cannot pass between these two places.

The horizon problem makes one wonder: If the universe is really so big that different parts of it are oblivious of each other, why is the cosmological principle valid? Why do all parts of a universe look the same?

8.3.3.8 Inflation and More Questions The inflation scenario answers the horizon problem. It says that initially the universe was so compact that all parts *could* communicate with each other. Then, a rapid expansion called inflation made the universe so big that different regions became isolated. Inflation appears to solve other questions raised by the big bang model (there are many). However, new questions arise. For example, it is not clear how inflation was started and why it has stopped. Perhaps the recently observed accelerated expansion associated with dark energy is an indication that inflation is not just a phenomenon of the distant past.

8.4 The Meaning of It All

Relativity stretches our imagination and asks us to abandon common sense. It distorts and mixes space and time in ways we never would have imagined. If things as basic as space and time are incomprehensibly muddled, what can be trusted? The cosmology following from general relativity leads to additional mind-bending assertions. Is it really true that everything we see is insignificant compared to the dark matter we cannot see? Inflation seems like a fantastic fix of a flawed big bang model. Despite its strangeness, the big bang with inflation boasts many successes. In addition to the microwave background, it explains nuclear abundances and more.

Problems remain, and revolutionary surprises surely lie ahead because physics has not yet answered some basic questions. One glaring example is: What happened to the antimatter? There is much more to learn.

9

NUCLEUS

9.1 Introduction

The uranium nucleus packs 92 protons into a sphere 10,000 times smaller than the atom. At such short distances, the Coulomb repulsion of the protons is much greater than atomic forces and energies. There is an obvious question: Why doesn't the uranium nucleus explode and send all 92 protons flying? An attempt to answer this question is postponed to Chapter 10 on particle physics. The following describes nuclei and radioactivity, but it does not explain the force that holds the nucleus together.

9.2 Nuclear Properties

9.2.1 Size

All nuclei are tightly bound balls of protons and neutrons with a typical size on the order of the mathematically convenient distance called the *classical electron radius*, $r_e = 2.92 \times 10^{-15}$ meters. This radius has a simple definition. The energy needed to push two electrons to within one classical electron radius is the same as the electron's relativistic mass energy mc^2. This nuclear characteristic size is smaller than the Bohr radius by a factor of $(137)^2$. If an atom was expanded to 100 meters (roughly the length of a football field), the nucleus would be about 1 centimeter across.

9.2.2 Mass

Both neutrons and protons have about 1800 times the mass of an electron. Because nearly all the mass of an atom is contained in its tiny core, the electrons dance around a nucleus that barely responds. It is analogous to the sun's indifference to the motion of planets.

9.2.3 Isotopes

Two isotopes of a chemical element differ only in the number of neutrons in the nucleus. Hydrogen was the first discovered example. The nucleus of the most common hydrogen isotope is a single proton. The nucleus of the second hydrogen isotope, called the deuteron, contains both a proton and a neutron. One of every 6400 hydrogen nuclei is a deuteron. The extra neutron makes a deuterium atom about twice as massive as an ordinary hydrogen atom. However, the single electron for both isotopes means their chemical properties are nearly identical.

Ordinary water combines two hydrogen atoms with an oxygen atom. "Heavy water" replaces the ordinary hydrogen atoms with deuterium atoms. Light water and heavy water are similar, but not identical. For example, heavy water freezes at a temperature 3.8 K higher than ordinary water. Considering the chemical similarities of ordinary water and heavy water, it is a surprise that animals, including humans, cannot survive if more than half their water is replaced by heavy water. Biology is sensitive to small physical changes. Some have speculated that health differences result from small local variations in the concentration of heavy water. This is really a stretch.

Hydrogen is the element with the fewest protons. Uranium is the naturally occurring element with the most protons. The two most common isotopes of uranium are justifiably famous. The most abundant uranium nucleus contains 92 protons and 146 neutrons. It is called U-238 because 238 is the sum of the proton and neutron numbers. The U-235 nucleus, with three fewer neutrons, comprises only about 0.72% of naturally occurring uranium.

The chemical properties of the two uranium isotopes are almost indistinguishable, but the nuclear properties are different. U-235 is associated with nuclear reactors and atomic bombs. The bomb dropped by the United States on Hiroshima, Japan, in 1945 contained about 65 kilograms of uranium, roughly 80% of which was the relatively rare U-235. Concentrating the U-235 component of uranium (enrichment) is by far the most tedious and time-consuming aspect of atomic bomb manufacture. The most common enrichment process follows these steps:

1. Use chemistry to make uranium-hexafluoride (one uranium atom bonded to six fluorine atoms). This compound is a gas.
2. Feed the uranium hexafluoride into a centrifuge and spin it as fast as possible. The slightly heavier U-238 is thrown to the outside of the centrifuge with a little more force than the U-235, leaving more of the lighter isotope near the center.
3. Collect the slightly concentrated U-235 from the center region of the centrifuge.
4. Repeat this process over and over, using many centrifuges.

It is no surprise that the details of uranium enrichment have been closely guarded secrets. It is also not a surprise that many countries have gained access to these secrets through mysterious (or not so mysterious) means. The US invasion of Iraq in 2003 was in part justified by reports that Iraq was acquiring aluminum tubes that could be used as centrifuges for uranium enrichment and eventual bomb production. Most people with technical knowledge about uranium enrichment believed this claim was nonsense. But, the technical experts who knew the most about uranium enrichment usually held security positions: They were sworn to secrecy, so the myth of "weapons of mass destruction" persisted for some time.

9.2.4 Energies

The atomic energies holding atoms together are a few electron volts, where 1 electron volt $= 1.6 \times 10^{-19}$ joules. The nuclear energies that confine the protons and neutrons within a nucleus are typically a million times the atomic energies. This "strong force" binding the protons and neutrons is very different from the forces of gravity and electromagnetism. The inverse square law does not apply, and this force is effective only for short distances. The electrostatic repulsion of the protons opposes the nuclear attraction, but the strong nuclear force can overwhelm electrostatics within the nucleus. As the number of protons increases, the increased electrostatic force and the larger nuclear size make the nucleus unstable. Thus, no naturally occurring nucleus has more than 92 protons. The completely stable nucleus with the largest number of protons (82) is lead.

The factor of about 1 million between atomic energies and nuclear energies is the reason that nuclear power and nuclear bombs produce so much more energy than chemistry (burning oil or coal). Enriched uranium made mostly of U-235 produces about 83 million megajoules for each kilogram of fuel. (One megajoule is 1 million joules.) Gasoline used to run a standard automobile produces no more than 50 megajoules/kilogram, which is more than a million times less energy. Gasoline stores about 50 times as much energy per kilogram as a lithium ion battery, which in turn stores several times the energy of a standard lead-acid car battery. For energy storage, nothing beats the nucleus, and a battery's relatively meager energy density is a major technological problem that frustrates efforts to develop "clean" energy.

Before the dangers of radioactivity became clear, people speculated about nuclear-powered cars of the future. These fantastic cars would be powered for their entire lifetime by just a small amount of nuclear fuel. No gas stations or recharging stations would be needed. The spectacular concentration of nuclear energy means one could just drive until the car fell apart.

9.3 Radioactivity

Alchemists of old (including Isaac Newton) searched in vain for the philosopher's stone to accomplish the "transmutation" of elements. Although Newton did not know it, some elements spontaneously transmute. Radioactive decay allows a nucleus to change its identity by lowering the total potential energy. For example, a uranium nucleus (the common U-238) lives for about 4.7 billion years (on average) before it changes to a thorium nucleus and a helium nucleus. The thorium is also unstable, and further decays follow. Eventually, one ends up with lead—not gold.

9.3.1 Alpha, Beta, and Gamma Radioactivity

The transmutation of uranium to thorium is one of many examples of alpha decay. In each case, the total potential energy is lowered by the ejection of an *alpha particle* or helium nucleus (two protons and two neutrons bound together) as shown in Figure 9.1.

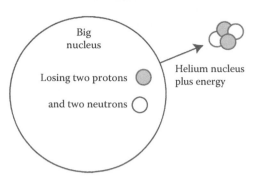

Figure 9.1   The ejection of an alpha particle from a large nucleus.

Beta decay is a different form of radioactivity that results in the ejection of an electron. The ejected electron is the beta particle. The physics of beta decay involves the "weak interaction" that is briefly described in Chapter 10. An important example of beta decay changes a carbon-14 nucleus to a nitrogen-14 nucleus.

Gamma decay occurs when a photon is created in a nuclear transition. The physics is analogous to the energy-conserving process that produces a photon when an atom makes a transition to a lower energy state. The photons associated with radioactivity (and high-energy photons in general) are called gamma rays. Gamma rays often accompany alpha and beta decay.

All three types of radioactivity can be dangerous, and gamma rays are especially hazardous because they travel long distances. Alpha and beta particles rapidly lose energy and do not travel far. Neutrons can also be produced by nuclear processes, and they can travel longer distances. Their lack of charge means they are not bounced about as much by surrounding electrons and nuclei. Neutrinos are nearly invisible particles that are also ejected in radioactive decays associated with the weak interaction. because neutrinos almost never interact with anything, they are perfectly harmless.

9.3.2 Alpha Decay and Tunneling

Alpha decay can only be explained using the quantum mechanical phenomenon of "tunneling." Classical physics fails to describe alpha decay because the alpha particle that is trapped inside the nucleus lacks the energy needed to overcome the barrier created by the strong nuclear attraction. However, quantum mechanics tells us the alpha particle (like the electron) has wave properties, and waves can leak though barriers. The alpha particle is like a prisoner repeatedly bouncing against the walls of a cell, wishing for a miracle. The miraculous quantum tunneling occurs only because the alpha particle hits the wall so frequently and because it is so small. As a rough estimate, assume the alpha particle moves at light speed divided by 137 and assume the nuclear diameter is twice the classical electron radius. Then, an alpha crashes into the nuclear boundary and attempts an escape about 3×10^{20} times every second. It almost always fails, but with this many tries, even the slightest possibility of tunneling through the barrier is significant.

An alpha's quantum wave extends a small distance into a forbidden region even though it does not have enough energy to surmount the barrier. In this never-never land of the classically impossible, the total energy is less than the classical potential energy.

Tunneling is illustrated in Figure 9.2. The wave describing the alpha particle approaching from the left decreases rapidly in the barrier region, and only a very small piece of the wave manages to sneak through the barrier. The probability of finding the alpha particle is proportional to the square of the wave function, so the chance of escape is small, but not quite zero.

This is a clear example of unpredictable quantum mechanics. One can calculate only the probability that the alpha will escape in each second. Our inability to determine the exact time of escape is not a result of sloppy mathematics or incomplete knowledge. The

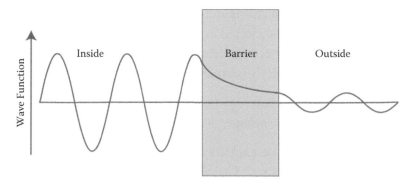

Figure 9.2 A particle incident from the left has a small probability of tunneling through the central barrier and appearing on the right.

uncertainty is built into quantum mechanics. The inability to make exact predictions is a *big* problem. As an analogy to Schrodinger's cat, is it possible that a nucleus has both decayed and not decayed?

Unpredictability for a single nucleus does not mean complete ignorance. Given a large number of radioactive nuclei, the *half-life* is the time needed for half of them to produce an alpha particle. Even though the decay time for a single nucleus is a mystery, the half-life of many identical nuclei can be precisely determined. The nuclei that failed to decay after one half-life are neither tired of waiting nor stubbornly sluggish. It takes exactly one more half-life for 50% of the remaining nuclei to decay, leaving 25% of them waiting. If one observes that only 1/16 survive, one knows that they originated four half-lives ago.

Radioactive alpha decay is not the only example of quantum tunneling. Tunnel diodes used in electronics and a tunneling scanning microscope are two of many practical applications of electron tunneling. Proton tunneling can explain the rare spontaneous mutations of DNA.

9.3.3 Beta Decay and Carbon-14 Dating

The half-life of radioactivity allows us to determine the ages of many materials. The best-known example is carbon-14 dating. The weak interaction of beta decay is the origin of this type of radioactivity.

Most carbon nuclei, called carbon-12, are stable. They contain six protons and six neutrons. Carbon-14 has two extra neutrons. This unusual form of the carbon nucleus is radioactive and undergoes beta decay. One of the neutrons in carbon-14 changes into a proton plus an ejected electron-antineutrino pair, as shown in the lower line of Figure 9.3. The unstable carbon-14 nucleus becomes a stable nitrogen nucleus with seven protons and seven neutrons. The half-life is 5730 years.

One might think that the instability of carbon-14 would mean these nuclei would have long ago disappeared. Nuclei were created billions of years ago, and that is many times the carbon-14 half-life. There is a trick shown in the upper line of Figure 9.3. High-energy particles called cosmic rays are constantly bombarding Earth. Some of these cosmic rays change a few of the nitrogen nuclei in our atmosphere into carbon-14 nuclei. An example path to carbon 14 is shown in the upper line of Figure 9.3, where an energetic neutron is absorbed by a nitrogen

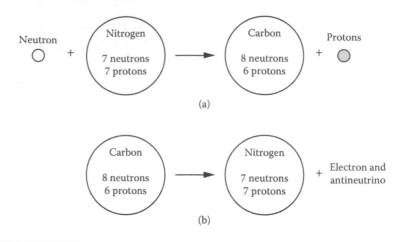

Figure 9.3 (a) Creation of carbon-14 by the absorption of an energetic cosmic ray neutron and ejection of a proton. (b) Radioactive beta decay of carbon-14.

nucleus and a proton is ejected. The continual creation processes mean the ratio of atmospheric carbon-14 to ordinary carbon is about one part in a trillion ($1/10^{12}$). This tiny fraction is not totally insignificant. Because 12 grams of carbon contain 6×10^{23} atoms, 1 gram of atmospheric carbon contains about 50 billion radioactive carbon-14 nuclei. Modern methods can determine the number of carbon-14 atoms, even in very small samples.

A famous example of carbon-14 dating gives the age of Dead Sea scrolls (ancient religious texts: Qumran Cave 1). The linen scrolls were made from the flax plant. As it grew, this plant incorporated a small amount of the atmospheric carbon-14 into its structure. As soon as the plant died, carbon absorption stopped. The fraction of carbon-14 atoms decreased in time as these unstable nuclei underwent beta decay. Measurement showed that carbon-14 concentration in the scroll material was only about three-quarters the carbon-14 concentration in the atmosphere. Because one-quarter was missing, the age is estimated to be about 2200 years. It is reasonable to assume the scrolls were written only shortly after the death of the plant, so we know (roughly) when these scrolls were written.

Even though there are some uncertainties and technical corrections, carbon-14 dating has progressed in accuracy since its origins around 1950. The Dead Sea scroll ages are reliably determined to within about 100 years.

Figure 9.4 A portion of the Shroud of Turin, photographed to enhance visibility.

The age of the Shroud of Turin, a portion of which is shown in Figure 9.4, is a more controversial example of carbon-14 dating. Some people believe Jesus was buried in this ancient cloth, but the carbon-14 analysis says the shroud dates from around 1300 years after the death of Christ. Historical records of the shroud prior to the year 1300 are murky. Some have suggested that the disagreement between a technical measurement and religious belief is a result of a scientific mistake. They suggest that the dated piece of the cloth was a much younger repaired fragment, but many experts are confident that the science is correct.

Carbon-14 is but one example of radiometric dating. The age of Earth can be estimated by comparing uranium and lead abundances. (Radioactive decays mean that uranium eventually changes into lead.) The complicated dating method contributes to our best-estimated Earth age of 4.5 billion years, or nearly one-third the age of the universe.

9.3.4 Gamma Decay and the Mossbauer Effect

A particularly important example of gamma decay is the 14,400-eV photon produced when an excited iron-57 nucleus makes a transition to a lower energy state of the same nucleus. Energy conservation precisely determines the energy of the photon (gamma).

Momentum conservation presents a complication. Just as a cannonball must share a little of the explosive energy with its recoiling cannon, not all of the nuclear energy is given to the gamma ray. The recoiling nucleus absorbs a small part of the energy because the photon has momentum even though it has no mass. The energy lost to the recoiling nucleus means the photon energy is not sufficient to excite another iron-57 nucleus.

When the iron nucleus is embedded in a solid, the physics is more complicated. Sometimes, the nucleus recoils and vibrates back and forth in the solid. But, there is also a chance that the recoil momentum is given to the entire solid. The large mass of the solid means the recoil is negligible, and a photon from one embedded iron nucleus can be absorbed by another nucleus. The second nucleus must also not recoil, so it also must be lodged in a solid.

This transfer of excitation energy from one nucleus to another (called a resonance) has important applications. The most famous example is the Pound–Rebka experiment (described in Chapter 8, Section 8.3.1) that tested general relativity. Gamma rays produced by the excited iron experience a small gravitational decrease in frequency as they travel up through Earth's gravity. The resonant absorption condition is thus destroyed. However, the resonance can be restored by moving the radiating and absorbing iron nuclei toward each other. Then, the Doppler shift increases the apparent frequency. A measure of the speed needed to restore the resonance determines the frequency shift.

This remarkably difficult verification of general relativity is a clear example of how different aspects of physics can combine to yield fundamental results.

1. General relativity supplies the basic idea.
2. Nuclear physics experiments yield the gamma ray.
3. Solid-state physics explains how recoil can be avoided.
4. The Doppler shift explains how resonance can be recovered.

9.4 Fission, Fusion, Nuclear Power, and Bombs

9.4.1 Fission

Some nuclei lower the total potential energy by breaking up into fairly large pieces. The most famous example is uranium-235. The breakup leads to two smaller nuclei, several free neutrons, and energetic gamma rays. An example (neglecting the gammas) is shown in Figure 9.5.

The key that makes this uranium nucleus both useful and dangerous is the chain reaction. Ejected neutrons can induce additional uranium nuclei to split. If not controlled, the fission multiplies so rapidly that an enormous amount of energy is released in a short time. The "atom bomb" is based on this chain reaction. Controlling the chain reaction by absorbing just the right fraction of the neutrons tames the atom bomb. It becomes a nuclear power plant.

9.4.2 Fusion

Fusion combines two nuclei to make a larger nucleus. Fusion powers the sun. The core of the sun is mostly hydrogen. Solar fusion is a several-step process that changes some protons into neutrons and then combines the two protons and two neutrons to form a helium nucleus. The potential energy of the helium is much lower than that of its constituent parts, so a great deal of energy is released.

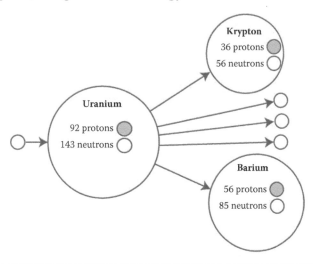

Figure 9.5   An example of nuclear fission with the production of two additional neutrons that can lead to a chain reaction.

The hydrogen bomb is a little taste of what is happening in the sun. The repulsion of protons is strong at nuclear distances, so to achieve fusion a large amount of energy must be supplied to overcome the Coulomb repulsion. In the sun, the energy is supplied by heat, and the process is facilitated by the enormous pressure of gravity. For a hydrogen bomb, the fusion is initiated by a nuclear fission bomb. The "small" atom bomb ignites the "big" hydrogen bomb.

9.4.3 Bombs and Ethics

The development of the atomic and hydrogen bombs is a dramatic example of the interplay between science and ethics. Most American scientists involved in the development of the atomic bomb believed it was essential because of the Nazi threat. There was concern that Germany would obtain a bomb first, with catastrophic consequences. Werner Heisenberg was the most famous German scientist associated with the German bomb. Details of his participation remain controversial.

After the defeat of Germany, a majority of the scientists involved in the bomb development favored an offshore demonstration, rather than the bombing of a Japanese city. They were not in control, and President Truman's decision was supported by the majority of the American public and the military. Robert Oppenheimer was in charge of the atomic bomb development, but his opposition to the development of the much more powerful hydrogen bomb combined with his leftist background led to his elimination from military research.

Of course, it is impossible to really know the ethical and practical considerations that guided scientists like Heisenberg and Oppenheimer to their decisions about what research is morally justified. This dilemma with ancient origins persists today. For example, some scientists refuse to participate in any project that requires security clearance. Others feel it is immoral not to assist in the national interest—even when that interest involves weapons.

9.4.4 Controlled Fusion

There is no obvious ethical dilemma associated with the development of controlled fusion. Having a power plant based on the physics of our sun is an appealing idea. In principle, controlled fusion could be

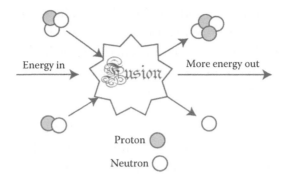

Figure 9.6 Fusion of a proton with one neutron and a proton and two neutrons yields a helium nucleus and an extra neutron. A great deal of energy is needed to push the protons close enough to fuse. When the fusion occurs, even more energy is released.

a source of nearly limitless energy. However, the practical problems are formidable—perhaps insurmountable. One proposed method of achieving controlled fusion heats protons bound to one or two neutrons, as shown in Figure 9.6. When these particles collide, some of the collisions may be so violent that the Coulomb repulsion between the protons will be overcome, fusion will occur, and a helium nucleus with a great deal of additional energy will be produced.

The considerable energy required for close approach of the protons means that anything containing the energetic particles would melt. In principle, a magnetic field in the shape of a donut, like that shown in Figure 4.20, can prevent the escape of the energetic charge particles. In practice, there are many problems with this "magnetic confinement." Another heating approach based on blasting a target with many powerful lasers presents equally challenging technical problems. Despite enormous effort and expense over many decades, no usable fusion energy has ever been obtained. Some are beginning to suspect too much optimism from the advocates of controlled fusion.

9.5 Nuclear Theory

The complexity of nuclear physics is clear when it is compared to atomic physics. The foundation of atomic physics is the hydrogen atom, consisting of one proton and one electron. The corresponding simplest two-body problem in nuclear physics is the deuteron, consisting of one proton and one neutron.

Quantum mechanics and Coulomb's law allow one to calculate the energy of the hydrogen atom as well as all the excitation energies. The 13.6-electron-volt binding energy of the hydrogen atom can be expressed in terms of the electron charge, the electron mass, and Planck's constant.

The deuteron's 2,200,000-electron-volt binding energy has no simple explanation. There are two reasons. First, the proton and the neutron are not fundamental point particles like the electron. They are made of "quarks." Second, the force between the constituent quarks is much more complicated than the electrostatic force of Coulomb's law. This strong nuclear force results from the interchange of "gluons." Gluons are not simple. Quarks and gluons are described briefly in Chapter 10.

Even though there is no simple nuclear analogue to the Coulomb force of atomic physics, approximate models characterize many aspects of nuclear structures and energies. An important example is the shell model analogous to the orbital model of atomic physics.

10
PARTICLE PHYSICS

10.1 Introduction

Particle physics asks deep questions. There are many. What are the basic building blocks of everything? What holds these blocks together to make the world we know? Where did they come from? Efforts to answer these questions have led to experimental searches requiring increasingly energetic particles and increasingly complex measurements. The experiments have been accompanied by theories that show elegance and unity. But, the job is not done.

10.2 Experiment and Theory

10.2.1 Experiments

Photons with energies of a few electron volts provided insight into atomic structure. X-rays with thousands of electron-volt energies probed matter. Rutherford's atomic model used alpha particles with nearly 5 million electron-volt energies. Higher-energy particles can probe the structure of the nucleus. The cyclotron particle accelerator pushed energies higher. Many improvements and alternative designs have followed. High energy is needed to create new, more massive particles because $E = mc^2$.

10.2.1.1 Accelerators and the Higgs Particle The highest-energy accelerator is the Large Hadron Collider at CERN (European Organization for Nuclear Research). This giant structure lies under the border between France and Switzerland. It accelerates two beams of protons on roughly circular paths with more than a 6-kilometer radius. Particles in each beam can achieve energies of several teraelectron volts (trillion or 10^{12} electron volts). The particles in the counterrotating beams collide. The energies from these collisions produce exotic short-lived particles never seen under ordinary situations.

Assigning the word *particle* to new high-energy phenomena is being a bit generous. An example is the Higgs particle. It lives (on average) for only a little more than 10^{-22} seconds. Traveling at 90% the speed of light, it would only move about 1% of the distance across an atom before it decayed into other short-lived particles. The Higgs mass is more than 100 times the mass of a proton. Thus, $E = mc^2$ means very high energy is needed for its creation.

After they are produced by a high-energy accelerator, detecting a particle like the Higgs is as challenging as making the particle in the first place. It is curious that the ATLAS detector used to find fundamental particles that are lighter than some atoms has a mass of more than 7000 tons. Charged particles can be seen because they interact with matter and leave a track. Uncharged particles, like the Higgs, do not. One should be spared the detail of how one can find an invisible particle that disappears almost immediately, but respect should be given the imaginative legions of scientists, engineers, and technicians (as well as administrators and accountants) needed to achieve the detection.

The discoveries of particle physics, including the Higgs particle, do not answer all questions of particle physics. New results inevitably lead to new mysteries whose answers may require even more energy. The Large Hadron Collider may not be up to the task. An accelerator producing even more energetic particles is conceivable, but at enormous cost.

The most energetic cosmic ray particles hitting Earth have energies millions of times the energy produced by the Large Hadron Collider, but no one has found a way to detect the exotic particles that are surely produced by the cosmic rays as they interact with the top of our atmosphere. By the time the particle shower has reached us, the shared energy (per particle) is no longer so impressive.

10.2.2 Theory

The path toward modern theories of particle physics is broadly traced as follows:

1. The first step into high-energy physics requires a relativistic extension of quantum mechanics. The Dirac equation provided a quantum description of a relativistic electron. The electron's antiparticle (the positron) followed naturally.
2. Special relativity tells us that not even forces can travel faster than light. Quantum electrodynamics (QED) explains this because electromagnetic forces are really the exchange of light-speed photons.
3. Generalizations of QED describe the forces associated with radioactivity and nuclear binding. The additional forces result from the exchange of "weak bosons" and "gluons" rather than photons. These extra force messengers (and much more) are unified by the "standard model." It provides a successful theoretical foundation of *almost* everything.
4. The obvious shortcoming of the standard model is its inability to unify quantum theory with gravity. String theory is one possible path to unification, but it is has been a "work in progress" for a long time. Supersymmetry is another extension of the standard model for which there is not yet any experimental verification.

The path from special relativity and the Dirac equation to QED and then the standard model is described in the next three sections.

10.3 Dirac Equation

According to special relativity, energy, mass m, and momentum p are related by the energy triangle of Chapter 8:

$$(Energy)^2 = (mc^2)^2 + (pc)^2.$$

Thinking mathematically rather than physically, there are two *Energy* solutions to this equation. One solution gives a sensible positive energy. The alternative negative energy seems wrong. Common sense says the negative energy particles should be ignored, but this "elephant in the room" will not go away. Quantum mechanics allows exotic processes like tunneling through barriers. Relativistic quantum theory makes the tunneling from a positive energy to negative possible, so new ideas are needed.

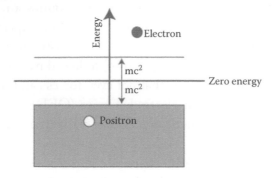

Figure 10.1 The electron energies are greater than mc^2, and the negative energies are less than $-mc^2$. The positron is a missing negative energy particle. Electron and positron annihilate each other when the electron falls into the empty negative energy hole. The extra energy is carried away by photons.

New ideas were supplied by Paul Dirac, who constructed a quantum-mechanical equation that describes relativistic electrons. Dirac's successful work could not avoid negative energy particles. He resolved this dilemma (after some delay and with advice from others) by noting that the *absence* of a negative energy electron is the same as the presence of a positive energy "antiparticle" called the positron. Dirac's positron makes sense because of Pauli's exclusion principle. When all negative energy levels are filled, there is room for no more. Removing one of the negative energy particles creates a positron and increases the total energy. Experiments discovered the positron 4 years after Dirac's prediction. It is an understatement to say Dirac's first-ever prediction of an antiparticle was an impressive accomplishment.

Modern treatments of positrons abandon the negative energy sea shown in Figure 10.1, but its intuitive appeal remains. An alternative and more magical interpretation depicts the positron as an electron moving backward in time.

There is a famous and unresolved question about the rarity of positrons. If the universe started with nothing but energy, charge conservation would mean equal numbers of positive and negative charges should be created. So, where are the positrons?

10.4 Quantum Electrodynamics

Quantum electrodynamics adds photons and their interactions to Dirac's electrons and positrons. A visual aid for QED is the Feynman diagram.

These diagrams help one to understand the basic ideas on a qualitative level. They also provide precise recipes for evaluating measurable quantities. For example, QED yields remarkably accurate numbers for atomic energies and radiation rates and for the scattering of electrons and photons.

10.4.1 Feynman Diagrams

A Feynman diagram consists of straight lines representing electrons or positrons, wiggly lines representing photons, and vertices corresponding to the creation or annihilation of a photon. The diagrams shown here are simplified (with an added time arrow) because only qualitative aspects of processes are considered.

The Feynman diagram of Figure 10.2 shows a photon exchange process associated with the Coulomb interaction. It is tempting to say electron-electron repulsion is similar to what happens when a ball is thrown from pitcher to catcher. The recoils of throwing and catching push the ballplayers apart. Perhaps the two electrons are like the pitcher and catcher and the photon is like the baseball. This is too much intuition. A similar Feynman diagram describes the electron-positron interaction, but they attract each other. When properly evaluated, the diagram of Figure 10.2 yields the Coulomb force with the correct sign for every case. Feynman diagrams with larger numbers of vertices also contribute to the force between electrons, but their corrections can often be ignored.

The Feynman diagram of Figure 10.3 (using the time arrow pointing up) shows electron-positron annihilation accompanied by the production of two photons. This is a detailed view of what happens when a positive energy electron falls into an unoccupied negative energy

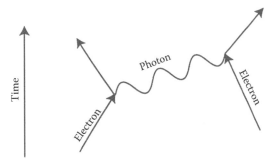

Figure 10.2 A Feynman diagram of electron-electron scattering. The solid lines represent electrons, and the wiggly line is a photon.

state, as suggested by Figure 10.1. In Feynman diagrams, a positron is represented as an electron with its arrow pointing in the negative time direction. Two photons must be created to conserve both energy and momentum. In all Feynman diagrams, the total energy and momentum are the same for the initial and final states of the system. The energies at intermediate stages are ambiguous.

Feynman diagrams reveal a variety of symmetries. One class of symmetries is illustrated in Figure 10.3, where the time arrow has been drawn in four different directions. These time choices correspond to four different physical processes:

1. Time flows up: Electron-positron annihilation occurs, resulting in two photons.
2. Time flows to the right: Electron absorbing and radiating a photon. A photon is scattered by an electron.
3. Time flows down: Two photons create an electron-positron pair.
4. Time flows to the left: A positron instead of an electron scatters a photon.

The symmetry means the rates of all four processes can be obtained from one basic calculation. Note that changing the flow of time changes the identity of the particles. Electrons are identified by arrows flowing with time and positrons have their arrow direction reversed.

Each vertex of a Feynman diagram is associated with the mysterious 1/137 (of Chapter 6, Section 6.6). This suggests that diagrams with addition vertices are relativity unimportant. Unfortunately, there

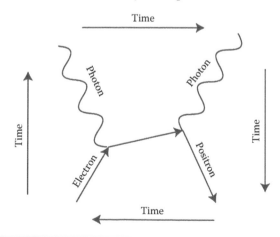

Figure 10.3   A Feynman diagram showing four processes, depending on the flow of time.

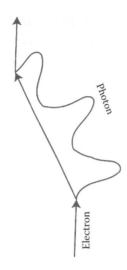

Figure 10.4 Feynman diagram showing an electron emitting and absorbing the same photon.

are many complications masked by a simple-looking Feynman diagram. Each diagram is only a guide to a complicated calculation of a number X that may be so large that $X/137^2$ in Figures 10.2 and 10.3 may not be so small. A bewildering example is shown in Figure 10.4, where an electron emits a photon and then absorbs the same photon.

The number X obtained from an evaluation of Figure 10.4 is infinite, so dividing by 137^2 is no help. This infinity arises because one must consider all the possible photons that could be radiated and absorbed. There are infinitely many because there is no lower limit on photon wavelengths. Clever mathematics and equally clever linguistic sleights of hand are needed to work around this apparent failure of the theory. When all is said and done, the infinities can be magically erased. QED appears to have no flaws when describing electrons, positrons, and photons. The elimination of infinities is called *renormalization*. Not all theories can be consistently renormalized, and this has been one of the stumbling blocks for the unification of quantum mechanics and gravity.

10.5 Beyond QED: The Standard Model

There is more to the world than electrons, positrons, and photons. One cannot ignore protons and neutrons, and there are many other objects called *particles*, even though they live for extremely short times and do not exist in ordinary matter. There are also additional forces beyond electromagnetism. These are described by the "standard model," which unifies most of physics. The standard model includes

1. All of QED.
2. The weak force and particles that transmit the weak force. An example is the W^-.
3. The "quark" components of particles like the proton and the colorful array of "gluons" that transmit the strong force between quarks.
4. Additional particles. Examples are the muon and the Higgs particle.

Almost anything that can be measured can be explained (in principle) by the standard model. However, even a list of the residents in the standard model's particle zoo is exhausting, and details are not presented here. The Feynman diagram of Figure 10.5 illustrates the weak force. This is followed by an even more superficial discussion of the strong force. No attempt is made to describe how the standard model incorporates all these interactions into a single framework.

10.5.1 Weak Interactions

10.5.1.1 Muon Decay A muon is essentially a heavy electron. The muon decays into an electron, neutrino, and antineutrino through weak interaction, as shown in Figure 10.5. Weak interactions are

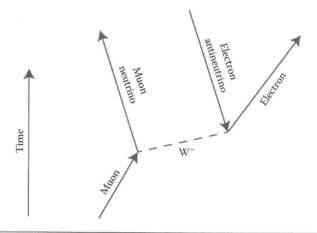

Figure 10.5 Decay of a muon into an electron, an antineutrino, and a neutrino. The massive W^- transfers energy, momentum, charge, and particle identities.

described by the interchange of "weak bosons" instead of photons. One of the exchanged particles is called the W^-. The W^- has the same charge as an electron and nearly 100 times the mass of a proton. The large mass keeps W^- from traveling far. As a result, the weak interactions have a short range. Like the photon, the W^- can both scatter particles and create particle pairs. Unlike the photon, the charge of the W^- means it can change the identity of particles in a more fundamental way.

10.5.1.2 Neutron Decay   A free neutron turns into a proton, an electron, and an antineutrino in about 15 minutes. The conservation laws obeyed by this neutron decay are illustrated in Figure 10.6. The number of heavy particles (neutron and proton), the number of light particles (electron and neutrino), the charge, the energy, and the momentum are all conserved. Energy conservation must include the rest-mass energy of the particles described by $E = mc^2$.

Neutron decay, like muon decay, involves the exchange of the W^- that obeys the conservation laws. However, neutron decay looks more complicated because neutrons and protons have internal structures. They are composed of three particles called quarks. Thus, neutrons and protons are drawn with three lines, one for each quark. The decay of a neutron, with its quark structure revealed, is shown in Figure 10.7.

The three-quark neutron is changed into a three-quark proton because W^- production changes the down quark (d) to the up quark (u). As with the muon decay, the W^- disappears and becomes an electron and an antineutrino.

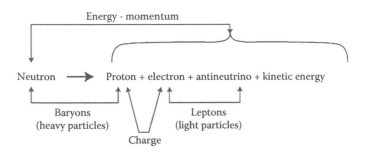

Figure 10.6   Neutron decay and associated conservation laws.

10.5.2 Strong Interactions

The neutron of Figures 10.7 and 10.8 is made of two down quarks and an up quark. Its charge is zero because the up quark has two-thirds the charge of a proton and the down quark has minus one-third this charge. Murray Gell-Mann's name *quark* was inspired by James Joyce (*Finnegan's Wake*). The neutron decay by the weak force into a proton, electron and anti-neutrino is shown in Figure 10.7. The quark structure of the neutron is illustrated in Figure 10.8.

One never sees the quarks that make up protons, neutrons, and other heavy particles. Another force particle, the gluon, holds the

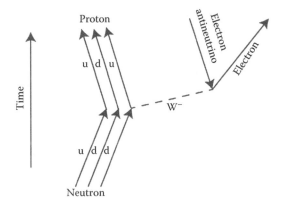

Figure 10.7   Neutron decay. The production of the *W*⁻ changes the d quark into a u quark.

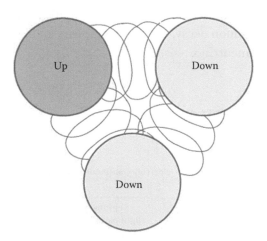

Figure 10.8   A neutron composed of two down quarks and an up quark is held together by the exchange of the spring-like gluons.

quarks together. The gluons are represented by the spring-like structures of Figure 10.8. Gluons are more complicated. They interact with each other as well as with the quarks. The further one separates two quarks in the neutron, the more the gluons will pull them back, making the observation of individual quarks impossible. In the nucleus, the gluons can be exchanged between different nuclei, and this is the aspect of the strong force that holds the nucleus together.

10.5.3 *There Is More*

The standard model has many more particles, including the famous Higgs particle that endows other particles with mass. The masses and interaction strengths of the standard model are determined by experiment, not by an overarching theory. There are more than 20 standard model numbers that have mysterious origins. The famous 137 is but one of these numbers. Many have noted that even a small change in these numbers could make the universe a far different place, generally a very uncomfortable one.

Some have attached theological significance to the fortuitous structure of the standard model. Perhaps God created the universe with exactly the correct fundamental constants to make it work. Then, God left the universe alone to develop as it pleased, without any further meddling. Albert Einstein may have favored this attitude at some times in his life.

Another explanation of the habitable universe is the "many-universe" picture. Perhaps our universe is not alone, and other universes have different values for the fundamental constants. Then, for the same reason that we live only on Earth and not Mars or the other planets, we inhabit only in the specially chosen universe with the comfortable fundamental constants.

10.6 The Future of Physics

10.6.1 *Applications and Technology*

Physics continues to make important contributions that make a difference to everyone's life. Examples include new materials, progress in fluid dynamics, miniaturization techniques, and quantum manipulations with applications to computing. The future may well include better batteries and solar cells that will make a big difference in our energy usage. However, the really exciting possibilities will be the ones we have not yet imagined.

10.6.2 Fundamental Questions: A Neutrino Example

The problems of particle physics hint at our unfinished search for the ultimate theory. The neutrino is a good illustration of mysteries yet to be solved. Its existence was postulated by Wolfgang Pauli long before this nearly invisible particle was "seen." Pauli's neutrino was necessary for energy conservation in radioactive decay. Zero-mass neutrinos are a component of the standard model. But, neutrino masses are not zero. One can ask many questions about neutrinos:

1. Why do the neutrinos have mass?
2. Why are the masses so small?
3. Why are there three kinds or neutrinos—or are there more?
4. Do neutrinos possess time reversal symmetry?
5. Can neutrino properties explain the lack of antimatter?

The mysteries of the neutrinos seem more dramatic when one realizes that they are the most abundant particles in the universe. They are as ubiquitous as photons and their total mass is comparable to the mass of everything else.

10.6.3 Fundamental Questions: A Broader View

1. There is something wrong with quantum mechanics. It predicts probabilities for experiments, not certainties. But, the formal mathematics of quantum mechanics predicts the quantum state with certainty. The mathematical conclusion is a Schrodinger's cat that is both alive and dead. Thus, those of us who observe Schrodinger's cat are also in a superposition of two states, but there is no way we can be aware of our other half. It impresses many as too exotic that we ourselves are in two (actually many) states and the probabilities of quantum

mechanics are only artifacts of our existence in the form of many quantum states.

2. String theory opened the door to a unification of gravity and quantum mechanics, but the door enters onto a mathematical jungle. Someone needs to open another door.
3. Supersymmetry might be the second door (or part of the first door), but so far experimental confirmation is lacking.

10.6.4 The Ultimate Answers

I do not share the faith of many scientists that the basic questions will eventually be answered. It may turn out that Eugene Wigner's idea is correct, and we are no more capable of solving the ultimate problems than a dog is capable of learning calculus. We may be barking up the wrong tree.

11

NOBEL LAUREATES

A Short Summary

11.1 Alfred Nobel

The merchant of death is dead.

This headline from Alfred Nobel's premature obituary pre-dated his death by 8 years. No doubt he was surprised. He was also upset. The atheist inventor of dynamite with pacifist feelings apparently felt uneasy about his reputation, and the Nobel Prize was the result. He had become enormously wealthy, amassing more than $200 million (in today's currency). The will that gave 94% of his fortune for the establishment of the Nobel Prize was written in secret. Not surprisingly, it was contested by relatives and questioned by authorities. Legal delays meant the first prizes were not awarded until 5 years after his death.

The medallion associated with the Nobel Prize is an 18-karat gold disk 66 millimeters in diameter with nearly pure gold on the surface (Figure 11.1). It is worth about $10,000.

Figure 11.1   The Nobel Prize includes a gold medal.

During World War II, there was a fear that the Nazis invading Denmark would steal the medals of Max von Laue (award date 1914) and James Franck (1925). A clever chemist, George de Hevesy, dissolved the medals in acid. In their liquid form, the medals were undetected. After the war, the gold was recovered and recast into its previous Nobel Prize form.

11.2 Recipients

The awards listed here are divided into several physics fields. The roughly chronological ordering of the works in each field shows how physics has progressed over more than a century. When appropriate, an approximate year the work was done is given following the description. The award year is given in parentheses, following either the description or the year the work was done. In some cases, the work pre-dated the award by many years (e.g., for Subrahmanyan Chandrasekhar).

11.2.1 Finding Particles

Discovering a new particle (or determining its fundamental properties) is a solid step toward a Nobel Prize. The science associated with particle discoveries has changed enormously over a century. The simple experiments of J.J. Thomson stand in striking contrast to the modern efforts of huge teams of scientists gathering and analyzing mountains of data generated by giant accelerators. In recent years, a person cited for a particle discovery is a team leader, not a solitary scientist working diligently in a secluded laboratory.

> *Lord Rayleigh (John William Strutt)* discovered argon, but he is famous today for much more, including the explanation of the blue sky. (1904)
>
> *Sir Joseph John Thomson* is generally recognized as having "discovered" the electron. 1897, (1906)
>
> *Robert Andrews Millikan* measured the electron charge. 1911 (1923)
>
> *James Chadwick* discovered the neutron. 1932 (1935)
>
> *Carl David Anderson* discovered the positron. 1932 (1936)

Emilio Gino Segre and *Owen Chamberlain* discovered the anti-proton. 1955 (1959)

Frederick Reines was the first to detect the neutrino. 1956 (1995)

Leon M. Lederman, Melvin Schwartz, and *Jack Steinberger* discovered the muon neutrino. 1962 (1988)

Burton Richter and *Samuel Chao Chung Ting* discovered the J/ψ particle, the first particle containing "charmed" quarks. 1974 (1976)

Martin L. Perl discovered the tau particle, the heaviest known lepton. 1977 (1995)

Carlo Rubbia and *Simon van der Meer* made decisive contributions leading to the discovery of the Z and W particles, which are the carriers of the weak force. 1983 (1984)

11.2.2 Predicting Particles

Successfully predicting a new particle is another promising path to a Nobel Prize.

Wolfgang Pauli predicted the neutrino, but his Nobel Prize was for the Pauli exclusion principle. 1931 (1945)

Paul Adrian Maurice Dirac's prediction of the positron was part of his work. 1931 (1933)

Hideki Yukawa predicted mesons. 1935 (1949)

Francois Englert and *Peter W. Higgs* predicted the Higgs boson. 1964 (2013)

The relatively recent prediction of the Higgs boson has a more complicated history than a Nobel Prize awarded to just two people would suggest.

11.2.3 X-rays

The first Nobel Prize in physics was for the discovery of x-rays. This curiosity that could not be explained by classical physics has become enormously important in materials physics, biology, and medicine.

Wilhelm Conrad Röntgen received the first Nobel Prize for his discovery of x-rays. 1895 (1901)

Charles Glover Barkla showed that each element produces its own characteristic x-rays. 1906 (1917)

Max von Laue discovered x-ray diffraction by crystals, making it clear that x-rays were waves and crystals were periodic. 1912 (1914)

Sir William Henry Bragg and *Sir William Lawrence Bragg* analyzed crystal structures using (William Lawrence) Bragg's law. 1912 (1915)

Karl Manne Geog Siegbahn related x-rays to atomic transitions with more precision than Charles Barkla. (1924)

11.2.4 Radioactivity and the Nucleus

The discovery and investigation of radioactivity started the path to nuclear physics.

Antoine Henri Becquerel discovered radioactivity. 1896 (1903)

Pierre and Marie Curie furthered Becquerel's research. (1903)

Enrico Fermi studied nuclear reactions and neutron bombardment. 1934 (1938)

Maria Goeppert-Meyer and *J. Hans D. Jensen* developed the shell model of the nucleus. 1949 (1963)

Aage Neils Bohr, Ben Roy Mottelson, and *Leo James Rainwater* considered collective motion in the nucleus. 1953 (1975)

Robert Hofstadter investigated nuclear structure through electron scattering. 1957 (1961)

Rudolf Ludwig Mossbauer found a path to precise measurements using nuclear radiation and absorption of gamma rays. 1957 (1961)

11.2.5 The Photon

This first step into quantum mechanics has evolved into an important diagnostic tool.

Wilhelm Wien showed that the most intense radiation frequency of a black body is proportional to its kelvin temperature. 1893 (1911)

Max Karl Ernst Ludwig Planck introduced the quantization of light's energy to explain the black body radiation spectrum, including Wein's law. 1900 (1918)

Philipp Eduard Anton von Lenard studied cathode rays (later called electrons) and observed the frequency dependence of the photoelectric effect. (1905)

Albert Einstein used quantized light energy (photons) to explain the photoelectric effect. 1905 (1921)

Arthur Holly Compton discovered the frequency shift of scattered light (Compton scattering), which is a consequence of photon energy quantization. 1923 (1927)

Kai M. Siegbahn improved x-ray photoelectron spectroscopy (XPS), which allowed identification of elements from the ejected electron energies. 1954 (1981)

11.2.6 Particles as Waves

Prince Louis-Victor Pierre Raymond de Broglie said particles are also waves. 1924 (1929)

Clinton Joseph Davisson and *George Paget Thompson* observed the wave properties of an electron. 1927 (1937)

11.2.7 Atoms

Hendrik Antoon Lorentz and *Pieter Zeeman* measured (Zeeman) and explained (Lorentz) the changing of atomic radiation frequencies produced by a magnetic field. 1896 (1902)

Jean Baptiste Perrin studied Brownian motion and showed that atoms really exist. (1926)

Ernest Rutherford produced a model of atomic structure, but his Nobel Prize in chemistry pre-dated his atomic model. 1911 (1908)

Niels Henrik David Bohr produced the first derivation of hydrogen atom properties. 1913 (1922)

Johannes Stark observed electric field effects on atomic radiation frequencies. 1913 (1919)

James Franck and *Gustav Ludwig Hertz* performed an experiment that confirmed Bohr's ideas. 1914 (1925)

Alfred Kastler related light, atomic spectra, and resonance. (1966)

Chandrasekhara Venkata Raman observed changes in scattered light frequency corresponding to atomic excitations. 1928 (1930)

11.2.8 Quantum Mechanics

Werner Karl Heisenberg was cited for "the creation of quantum mechanics." 1925 (1932)

Erwin Schrodinger's Nobel Prize was awarded because quantum theory based on the Schrodinger equation was "productive." 1926 (1933)

Max Born related probability to the quantum wave function. 1926 (1954)

Eugene Wigner applied symmetries to better understand quantum mechanics. (1963)

Roy J. Glauber developed a representation of quantum systems especially useful for lasers. 1963 (2005)

11.2.9 Spin and Magnetic Resonance

Otto Stern's award was for the Stern–Gerlach experiment, which demonstrated spin quantization. 1923 (1943)

Isidor Isaac Rabi developed a resonance method for determining the magnetic properties of atomic nuclei. 1937 (1944)

Felix Bloch and *Edward Mills Purcell* related particle spin, magnetic energy, and resonant frequencies. 1945 (1952)

11.2.10 Particles and Fields

Developments in the relativity modern area of physics regarding particles and fields have complicated histories. For example, many scientists contributed to the development of quantum electrodynamics besides Tomonaga, Schwinger, and Feynman.

Willis Eugene Lamb measured quantum electrodynamic corrections to atomic energies. 1947 (1955)

Polykarp Kusch determined quantum electrodynamic corrections to the electron's magnetic moment. (1955)

Sin-Itiro Tomonaga, Julian Schwinger, and *Richard Feynman* developed theoretical quantum electrodynamics. 1948 (1965)

Chen Ning Yang and *Tsung Dao Lee* correctly postulated physics that was not mirror symmetric. 1956 (1957)

Yoichiro Nambu used broken symmetry to describe the weak force. 1961 (2008)

James Cronin and *Val Fitch* investigated K-meson decay and found that nature is not rigorously time reversal invariant. 1964 (1980)

Makato Kobayashi and *Toshide Maskawa* applied broken symmetry to show that there are at least three quark families. 1973 (2008)

Sheldon Lee Glashow, Abdus Salam, and *Steven Weinberg* unified weak and electromagnetic interactions and described the weak neutral current and the Z boson. 1968 (1979)

Murray Gell-Mann organized particles and coined the terms *eightfold way* and *quark*. 1961 (1969)

Jerome I. Friedman, Henry W. Kendall, and *Richard E. Taylor* used scattering to study the quark model. 1969 (1990)

Gerardus t'Hooft and *Marinus J. G. Veltman* eliminated some of the infinities that plagued particle physics calculations. 1972 (1999)

David J. Gross, H. David Politzer, and *Frank Wilczek* discovered "asymptotic freedom," which avoids strong interaction problems at short distances and confines quarks at large distances. 1974 (2004)

11.2.11 *Experimental Innovations, Mostly for High-Energy Physics*

Albert Abraham Michelson measured the speed of light with extraordinary accuracy and showed it did not vary with relative motion. 1882 (1907)

Charles Thomas Rees Wilson invented the cloud chamber. 1910 (1927)

Walther Bothe developed a device that could simultaneously detect pairs of particles and applied it to a variety of physical phenomena. 1924 (1954)

Patrick Maynard Stuart Blackett developed a cloud chamber with many applications, including the study of cosmic rays. 1925 (1948)

John Douglas Cockcroft and *Ernest Thomas Sinton Walton* built a proton accelerator that produced nuclear reactions. 1932 (1951)

Ernest Orlando Lawrence invented the cyclotron. 1932 (1939)

Pavel Aleksevevich Cherenkov, Il'ia Mikhailovich Frank, and *Igor Yevgenyvich Tamm* discovered and explained the Cherenkov effect. 1937 (1958)

Cecil Frank Powell developed photographic emulsion particle tracking that led to the discovery of the pi meson. 1938 (1950)

Donald Arthur Glaser invented the bubble chamber for particle detection. 1952 (1960)

Luis Walter Alvarez improved bubble chambers and other particle detection methods, which led to many discoveries. (1968)

Georges Charpak developed multiwire proportional counter-particle detectors. 1968 (1992)

11.2.12 Astrophysics

Victor Franz Hess discovered cosmic rays. 1913 (1936)

Sir Edward Victor Appleton discovered the Appleton layer, an ionized region hundreds of kilometers aloft that reflects radio waves. 1926 (1947)

Subrahmanyan Chandrasekhar described the evolution of stars. 1930 (1983)

Hans Albrecht Bethe described the energy production in stars. 1939 (1967)

Hannes Alfven discovered the Afven wave in magnetic plasmas. 1942 (1970)

William Alfred Fowler described the stellar creation of the elements. 1957 (1983)

Arno Allan Penzias and *Robert Woodrow Wilson* identified the cosmic microwave radiation. 1965 (1978)

Sir Martin Ryle observed distant galaxies with improved radio telescopes. (1974)

Antony Hewish played a decisive role in pulsar discovery. 1967 (1974)

Russell A. Hulse and *Joseph H. Taylor, Jr.* discovered a new kind of pulsar that provided a check of general relativity. 1974 (1993)

Raymond Davis Jr. and *Mastoshi Koshiba* detected neutrinos from the sun and from supernovas. (2002)

Riccardo Giacconi contributed to the discovery of x-ray sources. (2002)

John C. Mather and *George F. Smoot* found anisotropy in cosmic microwave radiation. 1992 (2006)

Saul Perlmutter, Brian P. Schmidt, and *Adam G. Riess* found that the universe expansion is accelerating. 1998 (2011)

11.2.13 Superconductors and Superfluids

Heike Kamerlingh Onnes liquefied helium and discovered superconductivity. 1911 (1913)

Pyotr Leonidovich Kapitsa was a codiscoverer of superfluid helium. 1937 (1978)

Lev Davidovich Landau developed a theory of superfluid helium. 1941 (1962)

John Bardeen, Leon Neil Cooper, and *John Robert Schrieffer* produced the basic theory of superconductivity. 1957 (1972)

Brian David Josephson, Leo Esaki, and *Ivar Giaever* investigated quantum effects in superconductors. (1973)

David M. Lee, Douglas D. Osheroff, and *Robert C. Richardson* discovered superfluid helium-3. 1972 (1996)

Alexei A. Abrikosov, Vitaly L. Ginsburg, and *Anthony J. Leggett* produced (over the years) theories of superconductors and superfluids. (2003)

J. Georg Bednorz and *Karl Alexander Muller* discovered high-temperature superconductors. 1986 (1987)

Eric A. Cornell, Wolfgang Ketterle, and *Carl E. Weiman* produced superfluid-like dilute gasses. 1995 (2001)

11.2.14 Devices and Inventions

Gabriel Lippmann used the interference of light to produce color photographs. 1891 (1908)

Owen Willans Richardson explained how charges fly from heated surfaces. 1910 (1928)

Charles Edouard Guillaume discovered the useful nickel-steel alloys invar and elinvar. 1896 (1920)

Nils Gustaf Dalen made brighter lights and timers for lighthouses and navigation buoys. (1912)

Guglielmo Marconi and *Karl Ferdinand Braun* were pioneers of wireless communication (telegraph). (1909)

Percy William Bridgman produced very high pressures that had many applications in materials science research. (1946)

Fits Zernike invented the phase contrast microscope. 1932 (1953)

Ernst Ruska made the first electron microscope. 1933 (1986)

Dennis Gabor invented the hologram. 1948 (1971)

William Bradford Shockley, John Bardeen, and *Walter Houser Brattain* invented the transistor. 1947 (1956)

Alfred Kastler developed optical pumping that was a step toward the laser and maser. 1953 (1966)

Charles Hard Townes, Nicolay Gennadiyevich Basov, and *Aleksandr Mikhailovich Prokhorov* worked on the laser and maser. (1964)

Charles Kuen Kao developed optical fibers. (2009)

Jack S. Kilby helped develop integrated circuits. 1958 (2000)

Norman F. Ramsey made an atomic clock based on the hydrogen maser. 1950 (1989)

Willard S. Boyle and *George E. Smith* invented the "charge-coupled device," which can move small charges from place to place. 1969 (2008)

Zhores I. Alferov and *Herbert Kroemer* developed "heterostructures" (matching dissimilar semiconductors) with applications for lasers and transistors. 1936 and later (2000)

Gerd Binnig and *Heinrich Rohrer* developed the scanning-tunneling microscope. 1981 (1986)

John L. Hall and *Theodor W. Hansch* developed the optical frequency comb that can link radio frequencies to optical frequencies. 1993 (2005)

Albert Fert and *Peter Grunberg* discovered giant magnetoresistance. 1998 (2007)

Isamu Akasaki, Hiroshi Amano and *Shuji Nakamura* invented and developed the blue light-emitting diode (2014)

11.2.15 Materials, Magnetism, and Phase Transitions

Johannes Diderik van der Waals described the liquid-to-gas phase transitions and showed the two phases are essentially similar. 1873 (1910)

Louis Eugene Felix Neel was the first to suggest antiferromagnetism. 1932 (1970)

Philip Warren Anderson, Sir Nevill Francis Mott, and *John Hasbrouck van Vleck* made numerous contributions to the theory of magnetic and disordered systems over many years. (1981)

Bertram N. Brockhouse and *Clifford G. Shull* used neutrons to study solid structures. 1950 (1994)

Nicolaas Bloembergen and *Arthur Leonard Schawlow* used lasers to study materials. (1981)

Pierre-Gilles de Gennes improved theories of liquid crystals and polymers. (1991)

Kenneth G. Wilson revolutionized the theory of phase transitions. 1971 (1982)

Andre Geim and *Konstantin Novoselov* did basic work on two-dimensional graphene. 2004 (2010)

11.2.16 Exotic Quantum Mechanics and Small Cold Systems

Hans G. Dehmelt and *Wolfgang Paul* developed the ion trap. 1959 (1989)

Klaus von Klitzing discovered the quantized Hall effect. 1980 (1985)

Steven Chu, Claude Cohen-Tannoudji, and *William D. Phillips* used lasers to trap and cool atoms. (1997)

Robert B. Laughlin, Horst L. Stormer, and *Daniel C. Tsui* discovered and explained a quantum fluid with fractionally charged excitations. 1982 (1998)

Serge Haroche and *David J. Wineland* manipulated quantum systems. (2012)

11.3 Some Curiosities

Every recipient of a Nobel Prize was (or is) an outstanding scientist, but no ranking or prize can be completely objective and fair.

Deserving scientists have been overlooked, and there are curiosities associated with the choices of some recipients.

Ernest Rutherford is an interesting example. His model of the atom, with a heavy nucleus at the center, is one of the most important steps toward modern physics. But, he did not receive a Noble Prize in physics. Instead, he was awarded the 1908 Nobel Prize in chemistry, 3 years before his groundbreaking atomic model. His chemistry prize is a bit ironic because of his famous quotation: "All science is either physics or stamp collecting." (Hans Bethe [1967 Nobel Prize] was a stamp collector.)

Another Nobel Prize in chemistry (rather than physics) was awarded to Willard Libby in 1960 for the development of carbon-14 dating. The key to the dating technique involves the physics of nuclei, radioactivity, and half-lives.

A third Nobel Prize in chemistry associated with physics was awarded to Otto Hahn in 1945 for the discovery of nuclear fission. Many believe that Lise Meitner should have shared in this prize. Several other people were also involved in the discovery of fission.

A controversial example is the Nobel Prize associated with the discovery of the first pulsar. A pulsar is a magnetic, compact, rotating star. The periodicity of the signal corresponds to the star's rotation rate. The first pulsar was discovered by Jocelyn Bell Burnell, who was a doctoral student. Her adviser was Antony Hewish (1974). The Nobel Prize was awarded only to Dr. Hewish "for his decisive role in the discovery of pulsars." Burnell is generous and believes the award was appropriate.

Arnold Sommerfeld made many contributions fundamental to physics, but he did not receive a Nobel Prize. Some think he was overlooked because of anti-German prejudice following World War I.

Enrico Fermi's Noble Prize (1938) cites an accomplishment that turned out to be a mistake. He thought he had created new heavier nuclei through neutron bombardment. The created particles were actually fission products. It is ironic that Fermi did not realize he had produced fission a few years earlier than the work by Otto Hahn that clearly established fission.

There is no doubt that Fermi deserved a Nobel Prize for something. His breadth of achievements in both experimental and theoretical physics is amazing. He led the research to develop the first nuclear reactor. He made key theoretical contributions to statistical physics, beta decay, and cosmic rays. Fermi–Dirac statistics, fermions, and fermium are aptly named. Fermi also named the neutrino.

Many feel that E. C. G. Sudarshan was unfairly overlooked *twice*, having done early work on the unification electromagnetic and weak interactions (1979 award to Sheldon Glashow, Steven Weinberg, and Abdus Salam) and quantum representations of coherent light (2005 award to Roy J. Glauber). The 2005 award is considered by many to be a serious oversight. A number of scientists sent complaints to the Swedish Academy. In 2007, Sudarshan told the *Hindustan Times*, "The 2005 Nobel Prize for physics was awarded for my work, but I wasn't the one to get it."

The cosmic microwave background radiation remnant of the "big bang" was predicted by Ralph Alpher and Robert Herman in 1948, which was 17 years before it was accidentally discovered by two careful and thorough scientists, Arno Penzias and Robert Wilson (1978). The 1948 prediction and the 1965 discovery are only part of a complex history involving many people over many years. This Nobel Prize is an example of being at the right place at the right time. Penzias and Wilson were not looking for signs of the big bang, but they had access to the people who were aware of the significance of their discovery.

When finances are involved, controversies can become public. The Nobel Prizes for the telegraph and the laser are two examples.

Many have argued that Nicola Tesla was as deserving (or more deserving) of a Nobel Prize than Guglielmo Marconi, who shared the 1909 Nobel Prize with Karl Braun. Tesla's many passionate supporters may have a point, but the development of radio and telegraph is much more complicated than a disagreement between two inventors. Marconi borrowed ideas from many and was not generous in sharing credit. However, he did manage to get things done. His political and business skills as well as his family's prosperity allowed him to achieve many firsts in wireless transmission. The claim that the Nobel committee considered awarding a Noble Prize to Nicola Tesla and Thomas Edison appears to be only a rumor.

By 1902, Marconi and associates had established wireless transmission across the Atlantic. Radio came of age just 40 years after Maxwell's equations first described electromagnetism. Heinrich Hertz was the first to demonstrate radio waves (and verify Maxwell's theory) in the 1880s. From a fundamental physics viewpoint, Hertz's work was much more important than Marconi's transmission. Maxwell may have had some insight into the eventual applications of his theory. When asked about the use of electricity, he replied: "I do not know, but I'm pretty sure Her majesty's government will soon tax it."

Controversy surrounding the invention of a laser developed into a long-lived lawsuit. Many fine scientists contributed to laser development. It is not clear that the legal challenge clarified the question regarding who did what first. It is clear, however, that Charles Townes (1964) was the primary developer of the maser. A maser amplifies microwave radiation instead of visible light.

Another lawsuit was filed by Oreste Piccioni against Owen Chamberlain and Emilio Segre (1959). Piccioni was troubled by the lack of recognition he received for his role in the discovery of the antiproton.

Albert Einstein's 1921 Nobel Prize is probably the best-known curiosity. Several years after his revolutionary work on special relativity and general relativity, his Nobel Prize citation failed to mention relativity. After his remarkable accomplishments in 1905, it was certain that Einstein would win a Nobel Prize. Its anticipated value was part of his 1919 divorce settlement with Mileva Maric.

11.4 Traits of Nobel Prize Winners

There is no sure path to a Nobel Prize.

11.4.1 Intelligence

No fools have won a Nobel Prize (at least not in physics), but intelligence is a little vague.

Richard Feynman (1965) seemed amused that his measured high school IQ was 125. This is pretty good, but not spectacular. An IQ of 125 means more than 280 million people alive today are smarter than was Richard Feynman. Where are they? George F. Smoot (2006) won $1 million on the TV quiz show *Are You Smarter than a Fifth Grader?* so his IQ was probably at least 125. Some Nobel Prize winners have been prodigies, but Albert Einstein (1921) is one of a few examples who appeared to be slow at an early age. Wolfgang Pauli was as smart as they come, and he knew it. One of his favorite criticisms of insignificant work was: "It is not even wrong."

Many Nobel Prize winners were great teachers. Enrico Fermi (1938) and Erwin Schrodinger (1933) are examples. Others had more difficulties. Eugene Wigner (1963) struggled to find the level of his audience. Isidor Rabi (1934) and Max Planck (1918) were described by some as "the worst teacher ever," but opinions varied. Despite this, the accomplishments of the less-than-charismatic teachers were sufficient to inspire many younger scientists.

11.4.2 Family Ties

Nobel Prize winners Aage Niels Bohr (1975), William Lawrence Bragg (1915), Kai Siegbahn (1981), and George Paget Thomson (1937) were the sons of Nobel Prize winners. Subrahmanyan Chandrasekhar (1983) was the nephew of C. V. Raman (1930). Pierre and Marie Curie's daughter, Irène Joliot-Curie and her husband, Frédéric Joliot, received a Nobel Prize in chemistry (1935).

The academic environment of these families was surely a major influence. Many other Nobel Prize winners benefitted from stimulating intellectual home environments and distinguished academic families. Three examples are Max Planck (1918) whose father, grandfather and great-grandfather were professors, Arthur Compton whose father was a dean and professor of philosophy, and Antoine Becquerel whose grandfather, father and son were all scientists. Nobel Prize winners who did not have the benefit of an academic family often had their gifts recognized at an early age and were afforded the education that appears to be indispensable for significant achievement in physics.

11.4.3 Collaboration

Famous collaboration examples are those of Pierre and Marie Curie (1903) and William Lawrence and William Henry Bragg (1915). The husband-wife pair and the father-son pair shared Nobel Prizes. There are many other cases for which an accomplishment was a team effort. In modern experimental physics, the team can comprise a large number of people. Singling out a special few persons for recognition can be difficult and controversial, but Nobel Prize tradition limits the number of recipients to three.

Some collaboration can be difficult. The case of William Shockley and the 1956 Nobel Prize for the invention of the transistor is a famous example. It would be an understatement to call William Shockley "difficult." John Bardeen and Walter Brattain (corecipients for the Nobel Prize) soon separated from Shockley. His stubborn ego and paranoid tendency made working with others increasingly difficult. He was eased out of industrial leadership. His isolation was furthered by his outspoken and unpopular views on societal ideas. In particular, he was concerned about "dysgenics" (the lowering of average intelligence by the higher reproduction rate of dumb people). By the time of his death, few loyal family members remained, and the "friends of Shockley" club was small. This was a sad end for a gifted scientist.

11.4.4 Isolation

Collaboration is not always the key, and sometimes the accomplishment seems quite solitary. This appeared to be the case for Einstein's special relativity, despite claims that his first wife, Mileva, made substantial contributions. Another example is Louis de Broglie, who said: "After long reflection in solitude and meditation, I suddenly had the idea, during the year 1923, that the discovery made by Einstein in 1905 should be generalized by extending it to all material particles and notably to electrons."

11.4.5 Religion and Philosophy

Richard Feynman's skeptical attitude toward the philosophy of science has been shared by many hard-headed physicists.

> Philosophy of science is about as useful to scientists as ornithology is to birds. (attributed to Richard Feynman)

In a broader sense, philosophy can have enormous influence on science because it affects the attitude of society in general. The culture of the Middle Ages was not helpful to science, as evidenced by Galileo's problems. A more striking example is Giordano Bruno, who speculated that the sun was just another star. For this and other "errors," he was burned at the stake in 1600.

More recently, nineteenth-century empiricism rejected physical objects that could not be observed, and this included atoms and molecules. This attitude made it especially difficult for Ludwig Boltzmann to gain acceptance for his pioneering ideas of the atomic theory of statistical physics. It is not clear to what extent this contributed to Boltzmann's suicide.

Today, philosophy colors our opinions about quantum mechanics and the meaning of probability. Speculations about Schrodinger's cat and multiple universes are colored by our views of common sense and free will.

11.4.5.1 Does One Need to Be an Atheist to Win the Nobel Prize? Religious faith, especially of the traditional type, is surely less common among the Nobel Prize winners. A list of the nonreligious, the agnostics, and the atheists (in physics) probably constitutes a majority of individuals. It is hard to be certain because religion is a private matter for many, and a person's religious views often change with age. But, some public nonreligious Nobel Prize winners are well known. They include Philip Anderson (1977), John Bardeen (1956 and 1972), Hans Bethe (1967), Niels Bohr (1922), Subramahyan Chandrasekher (1983), Pierre and Marie Curie (1903), Louis de Broglie (1929), P. A. M. Dirac (1933), Albert Einstein (1921), Enrico Fermi (1938), Richard Feynman (1965), Lev Landau (1962), Isidor Rabi (1934), C. V. Raman (1930), William Shockley (1956), Erwin Schrodinger (1933), Johannes Diderik van der Waals (1919), Eugene Wigner (1963), and Stephen Weinberg (1979). In this group, opinions varied. Enrico Fermi was an agnostic. At the other extreme, Steven Weinberg said "Religion is an insult to human dignity." Albert Einstein is an ambiguous case. He frequently mentioned God, but it is not obvious what he meant. Some have suggested that he thought God created the universe, but then left it alone.

The *Humanist Manifesto III* was signed by six physics Nobel Prize winners: Owen Chamberlain (1959), Philip W. Anderson (1977), Sheldon Glashow (1979), Jerome I. Friedman (1990), Pierre-Gilles de Gennes (1991), and David J. Gross (2004). This manifesto says science reveals the world and makes our lives better, evolution is valid, and people who help other people are happier. There is no mention of God in the *Humanist Manifesto.*

The religious Nobel Prize winners are an interesting minority.

Wolfgang Pauli (1945) is an example of the pitfalls of a simple description. His extended interactions with psychologist Carl Jung led him far from his Catholic background, but it is hard to say what religion meant to him.

Werner Heisenberg (1932) said: "The first gulp from the glass of natural sciences will turn you into an atheist, but at the bottom of the glass God is waiting for you." Others, including Charles Glover Barkla (1917), Max Born (1954), William Henry Bragg (1915), Robert Millikan (1920), and Max Planck (1918) held fairly traditional religious views, as does Charles Townes (1964), whose age at this writing was 99. Lev Landau (1962) chided James Franck (1925) on his religion. Landau thought it was outmoded for a scientist.

Brian Josephson (1973) is director of the Mind-Matter Unification Project at the University of Cambridge in England. He and his fellow workers study the relations between quantum mechanics and the human conscience. He believes physics may explain telepathy and psychokinesis. Some of his work is related to the teaching of Maharishi Mahesh Yogi. It is not clear if Josephson's ideas should be considered a religion. Pierre Curie also had an interest in the paranormal. He attended séances and took notes. He wondered if there was some connection between science and the spirit world.

Abdus Salam (1979) was a devoutly religious Ahmadiyya Muslim who claimed that all of science is part of Islam. The Pakistan government said his faith was not really Muslim. This is one reason he left Pakistan. Abdus had two wives and two separate families (up to four wives is acceptable). He married his cousin at age 23. Nineteen years later, he married Louise Napier Johnson, who was an Oxford professor. Both wives attended his Nobel award ceremony, which led to some amusing adjustments of standard protocol.

11.4.6 Prosperity

Some Nobel Prize winners had no trouble obtaining first-rate education and employment.

Louis-Victor Pierre Raymond de Broglie (1929) was born to a rich and aristocratic family. He obtained the best of French education at the Sorbonne and the University of Paris. Even employment was not a problem for him. After the First World War, he worked in his brother's physics lab. After the death of his brother, he became both a French duke and a German prince.

Antoine Henri Becquerel was also born into a distinguished French family. His father and grandfather were distinguished scientists.

John William Strutt (1906) became the Third Baron Rayleigh at age 31. His life was not a financial struggle, although there was some problem with the agricultural depression of the 1870s. His title included holdings of 7000 acres.

Guglielmo Marconi, First Marquis of Marconi (1909), was part of the aristocracy. The development of wireless communication was a race. An advantaged financial, business, and social background certainly helped Marconi win the race.

11.4.7 Hardship

Most Nobel scientists were not afforded the privileges Prince de Broglie or Baron Rayleigh enjoyed. It was typical that the academic tools and employment needed for scientific discovery were achieved through persistence and personal sacrifice. Some examples are extreme.

Marie Curie's (1903) ability to overcome hardship is legendary. She was discouraged as a student and frustrated in her attempts to obtain employment. Financial problems slowed her progress. Despite her many contributions, extra effort was needed to add her name to her husband's for their Nobel Prize. Only a dedicated and determined personality could have overcome the many obstacles she faced.

Abdus Salam (1970) was not treated with respect when he worked in Pakistan. He was assigned the job of soccer coach at the University of Punjab, and he was informed that soccer success was more important than his physics research.

Ernest Rutherford (1908, chemistry) was one of 12 children born on a New Zealand farm. Finances were tight, and Ernest remained frugal throughout his life. One of his many famous pronouncements was: "We haven't the money, so we've got to think."

George Charpak (1992) was imprisoned in Dachau during the Second World War. Walter Bothe (1954) was taken prisoner in the First World War and spent a year in Siberia. James Chadwick had the bad luck of working in Germany at the beginning of World War I. He was a guest of the Ruhleben Internment Camp for 4 years. Peter Kapitza (1978) was permanently detained on orders from Stalin when he attended a scientific conference in Russia.

Nils Gustaf Dalen (1912) was blinded while experimenting with acetylene gas just before he was to receive the Nobel Prize. Victor Franz Hess (1936) suffered a thumb amputation in 1934 because of an accident with radioactive materials. Marie Curie (1903) had cataract operations and suffered from lesions on her fingers and anemia because of her prolonged exposure to nuclear radiation. Even today, her cookbooks are kept in a lead box because of the remaining radioactivity.

Wilhelm Röntgen (1901) did not take patents out on his discoveries and donated his first-ever Nobel Prize money to the University of Wurzburg. The inflation after World War I led to his eventual bankruptcy.

11.4.8 Politics

Many physics Nobel Prize winners take on administrative roles related to science. Their service, especially in areas of nuclear energy, world peace, and education, is valued and appreciated. Others have broadened their activities. Zhores Alferov (2000) is an active participant in Russian politics. Stuart Blackett (1948) was a leftist active in the British Labor government politics. Steven Chu (1997) was secretary of energy during the first term of President Obama's administration.

The Nazi era was tragic for many reasons. Many scientists fled Germany because of their Jewish background or opposition to Nazism. These include Albert Einstein (1921), Hans Bethe (1967), Erwin Schrodinger (1933), Otto Stern (1943), Max Born (1954), and Felix Bloch (1952). Despite their important contributions to science, the reputations of Philip Lenard (1905) and Johannes Stark (1919) were destroyed by their Nazi sympathies. Max von Laue (1914) remained in Germany and bravely resisted Nazi ideas.

Max Planck (1918) was 74 years old at the time of the Nazi takeover in 1933. His attempts to moderate the government's actions and help Jewish scientists were admirable but not very productive. On a visit to Hitler, he said, "It would be self-mutilation to make valuable Jews emigrate, since we need their scientific work." Sadly, Max Planck's own son, Erwin, was killed by the Gestapo for his involvement in the 1944 plot to assassinate Hitler.

Werner Heisenberg's (1932) World War II role remains controversial. Most people believe that some aspects of his record are difficult to defend, but his descendants are among those who feel that he has been misrepresented. A famous meeting of Heisenberg and Niels Bohr (1922) is a subject of a great deal of speculation and a 2000 Tony Award-winning play, *Copenhagen,* by Michael Frayn.

In 1921, Albert Einstein accompanied Chaim Weizmann to the United States to encourage support for the establishment of a Jewish national homeland in Palestine. In 1939, Einstein signed a letter to President Franklin Roosevelt warning of the danger of a German atomic bomb.

Benito Mussolini was best man at Guglielmo Marconi's (1909) second wedding in 1927. Later in life, Marconi became a defender of the Italian Fascist regime and its invasion of Ethiopia. Marconi died in 1937 before all aspects of Fascism were apparent. Enrico Fermi left Italy in 1938 because of racial laws that affected his Jewish wife. Emilio Segre also left Italy in 1938 because of the laws barring Jewish employment. In 1943, Segre's father died in Rome, and his mother was captured by the Nazi regime. He never saw her again.

11.4.9 Music

An interest in music is not uncommon among Noble Prize winners. One cannot work on physics 24 hours a day.

Richard Feynman (1965) played bongos and picked locks. Max Born (1954), Donald Glaser (1960), Werner Heisenberg (1932), Gerard 't Hooft (1999), and Max Planck (1918) play(ed) the piano at various skill levels. Charles Glover Barkla (1917) and Frederick Reines (1995) were talented singers. Lord Rayleigh's (1904) interest in music led to the famous book *Theory of Sound* in 1877. George F. Smoot (2006) was a guest conductor of the University of California, Berkeley, marching band. Albert Einstein (1921) was a "relatively" good violin player but not a great violin player. An often-told story that may be true says a frustrated professional musician trying to accompany Einstein once said: "Albert, can't you count?"

Annotated Bibliography

Chapter 1. Newton and Mechanics

Wigner's very readable article is

Wigner, Eugene. 1960. "The unreasonable effectiveness of mathematics in the natural sciences." *Comments on Pure and Natural Sciences* Vol. 13, No. I. New York: Wiley.

The famous C.P. Snow lecture of 1959 is contained in the book that expands on his ideas:

Snow, Charles P. 2001. *The Two Cultures*. London: Cambridge University Press, p. 3.

A translation of Newton is

Newton, Isaac. 2010. *The Principia: Mathematical Principles of Natural Philosophy*. New York: Snowball.

One is surely impressed by this work, but reading it is not a good way to learn physics.

One of the foundations of modern science is

Galileo, Galilei. 1632. *Dialogue Concerning the Two Chief World Systems*.

Galileo showed imagination and brilliance in demonstrating constant acceleration. His rolling ball experiment is described in

Johnson, George. 2008. *The Ten Most Beautiful Experiments.* New York: Knopf.

This reference also describes experiments done by Newton (light), Faraday, Joule (heat and energy), and others.

The famous criticism of Aristotle and his lack of dental knowledge is described in

Russell, Bertrand. 1953. *Impact of Science on Society.* Brooklyn, NY: AMS Press.

One can overdo criticism of Aristotle, who has been called the first and last person to organize essentially all knowledge. In 1536, Petrus Ramus (logician and philosopher) allegedly claimed "everything that Aristotle had said was false." He was too unkind.

The complex intellectual developments leading up to Newton's work are described in

Rossi, Paola. 2001. *The Birth of Modern Science.* Cornwall, UK: Blackwell.

Feynman's famous comment on the importance of mathematics is in

Feynman, Richard. 1994. *The Character of Physical Law.* New York: Random House, Modern Library.

A translation of the Russian 1869 novel that makes an analogy with the inverse square law is

Tolstoy, Leo. 1994. *War and Peace.* Part 7, Chapter 1. New York: Random House.

The inverse square law is only approximate for objects in proximity like Nicolas and Sonya. Newton proved his results are exact for spheres. At the time, this proof was an impressive mathematical accomplishment.

Most modern measurements of the universal gravitational constant are not based on a revolutionary experimental technique. They reproduce the Cavendish experiment with greater precision. A survey is

Speake, Clive and Quinn, Terry. 2014. "The search for Newton's constant." *Physics Today* 67(7):27–33.

The authors pointed out that the torsion balance used by Cavendish was invented and constructed by John Michell, an overly modest and under-appreciated scientist.

The transit of Venus and associated adventures are colorfully described in

Anderson, Mark Kendall. 2012. *The Day the World Discovered the Sun.* Philadelphia: Da Capo Press.

Halley's hollow-Earth paper is

Halley, Edmond. 1692. "An account of the cause of the change of the variation of the magnetic needle; with an hypothesis of the structure of the internal parts of the Earth." *Philosophical Transactions of the Royal Society of London* 17:563–578.

Feynman's attempt to simplify Newton is described in

Goodstein, David and Judith. 1996. *Feynman's Lost Lecture: The Motion of Planets around the Sun.* New York: Norton.

A translation of Laplace's 1814 book that suggests a mechanical universe is

Laplace, Pierre Simon. 1902. *A Philosophical Essay on Probabilities.* New York: Wiley.

Edward Lorenz's published papers do not site the famous butterfly of the "butterfly effect." Seagull flapping is mentioned. His essential paper is

Lorenz, Edward. 1963. "Deterministic nonperiodic flow." *Journal of Atmospheric Sciences* 20:130–141.

A mathematical treatment of chaos is

Hubbard, John H. 2007. "The KAM Theorem." p. 215–238 in *Kolmovgorov's Heritage in Mathematics,* Editors Eric Charpentier, Annich Lesne and Nikolai Nikolski, Berlin: Springer-Verlag.

This article contains references to the most important original works. Despite an attempt at simplification, it is mathematically challenging.

A popular book that presents no mathematical challenges is

Gleick, James. 2008. *Chaos: The Making of a New Science.* New York: Penguin.

A collection of essays commenting on science, social science, and society is

Labinger, Jay and Collins, Harry (editors). 2001. *The One Culture?* Chicago: University of Chicago Press.

This work includes Steven Weinberg's view of the ultimate theory as well as his comments on the discovery of superconductivity.

Roger Penrose's famously sophisticated discussion of mechanistic science and its relation to the human conscience arrives at a conclusion very different from that of Laplace.

Penrose, Roger. 1989. *The Emperor's New Mind*. Oxford, UK: Oxford University Press.

Vera Rubin has written many articles on galaxy motion. An overview is

Rubin, Vera C. 1995. "A century of galaxy astronomy." *The Astrophysical Journal* 451:419.

It is curious that Ms. Rubin has been skeptical of the dark matter explanation of her own observations.

The first mention of dark matter (dunkle materie) is

Zwicky, F. 1933. "Die Rotverschiebung von extragalaktischen Nebeln." *Helvetica Physica Acta* 6:110–127.

Chapter 2. Momentum, Angular Momentum, and Energy

Da Vinci's parachute is drawn in notebooks with the description in the margin. The original notebooks are kept in Milan.

Da Vinci, Leonardo. c1483. Codex Atlanticus, Milan: Biblioteca Ambrosiana, folio 381v. Davic.

A physicist's discussion of the Kennedy assassination physics is

Alvarez, Louis. 1976. "A physicist examines the Kennedy assassination film." *American Journal of Physics* 44:813–837.

Kepler published several works that discussed his laws. The third law first appeared as

Kepler, Johannes. 1619. *Harmonices Mundi* [The Harmony of the World] Book 5 Chapter 3:189. Linz (Austria): Johann Planck.

The first two laws were refined in

Kepler, Johannes. 1622. *Epitome Astronomiae* Book 5 part 1. Linz (Austria): Johann Planck.

Insight into everyday physics, including excitation of the violin string resonance, is presented in

Varlamov, Andrey and Aslamazov, Lev. 2012. *The Wonders of Physics*. 3rd edition. Singapore: World Scientific.

Clocks before and after Galileo are described in

Landes, David S. 1983. *Revolution in Time*. Cambridge, MA: Belknap Press (Harvard).

Chapter 3. Thermal Physics

Strictly correct statements of the thermodynamics laws are worded more carefully than the simplifications presented in Chapter 3. Also, there is a "zeroth" law that says the almost obvious: "If two objects are at the same temperature as a third object, then the two objects have the same temperature."

The first accounts of Maxwell's demon are in letters dating from 1967. The work "demon" was introduced by Lord Kelvin in 1874. The quoted account is from Maxwell's book that first appeared in 1872.

Maxwell, James Clerk. 1902. *Theory of Heat*. London: Longmans, Green and Co.

Solid arguments for the view that renewable energy will take a long time to develop are presented in

Smil, Vaclav. 2014. "The long slow rise of solar and wind." *Scientific American* 310(1):52.

Lord Kelvin's characterization of Earth's age is described in

Burchfield, Joe D. 1990. *Lord Kelvin and the Age of the Earth*. Chicago: University of Chicago Press.

Darwin's first edition was

Darwin, Charles. 1859. *On the Origin of Species by Natural Selection, or the Preservation of Favoured Races in the Struggle for Life*. London: John Murray.

The title was shortened in later editions.

The earliest and most famous of Lyell's work are

Lyell, Charles. 1830, 1832, 1883. *Principles of Geology* 2, 1, 3. London: John Murray.

Mark Twain's suggestion that the most famous expert is most likely to be right is in

Twain, Mark. 2011. *Letters from the Earth*. Letter #ix. Seattle, WA: Pacific Publisher's Studio.

Chapter 4. Electromagnetism

The oil-drop experiment that measured the electric charge is described in

Millikan, R.A. 1913. "On the elementary electric charge and the avogadro constant." *Physical Review* 2:109–143.

There is controversy surrounding the ethics of Millikan's methods and authorship.

The charge-to-mass ratio is reported in

Thomson, J.J. 1897. "Cathode rays," *Philosophical Magazine* 44:293.

Einstein's groundbreaking paper was

Einstein, A. 1905. [Translated] "On the electrodynamics of moving bodies." *Annalen der Physik* 17:891–921.

The development of the foundations of electromagnetism are described in

Forbes, Nancy. 2014. *Faraday, Maxwell and the Electromagnetic Field: How Two Men Revolutionized Physics*. Amherst, NY: Prometheus Books.

The unfair Ohm's law criticism is described in

Hart, Ivor B. 1923. *Makers of Science*. London: Oxford University Press, p. 243.

Chapter 5. Waves

Two good sources on sound are

Goldsmith, Mike. 2012. *Discord; The Story of Noise*. Oxford, UK: Oxford University Press.
Robinson, Henry W. 2007. *Sound in the Atmosphere*. Port Orange, FL: Lochlyn Press.
Wolfe, Tom. 1979. *The Right Stuff*. New York: Farrar, Strauss and Giroux.

The first is less complicated.

Young's fundamental demonstration of the wave nature of light is found in

Young, Thomas. 1804. "Experimental demonstration of the general law of the interference of light." *Philosophical Transactions of the Royal Society of London*, 94: 1–16.

Young presented his ideas before the date of his experiment, and some have speculated that he may not have preformed his famous "double-slit experiment." Young's paper can be found in

Shamos, Morris (editor). 1959. *Great Experiments in Physics*. New York: Holt Reinhart and Winston, pages 96–101.

Doppler's paper "On the coloured light of binary stars and some other stars of the heavens—Attempt at a general theory including Bradley's theorem as an integral part" was independently printed in 1842. It was reprinted in 1843 in the *Proceedings of the Bohemian Society of Sciences*.

A very fine and famous book describing many everyday aspects of light is

Minnaert, Marcel. 1993. *Light and Color in the Outdoors*. New York: Springer-Verlag.

A description of the expanding universe and the history of its discovery is in

Livio, Mario and Riess, Adam. 2013. "Measuring the Hubble constant." *Physics Today* 66(10):41.

Chapter 6. Quantum

Max Planck first described his blackbody radiation law at the meeting of the German Physical Society (DPG) in1900. The subsequent publication was

Planck, M. 1901. "On the law of distribution of energy in the normal spectrum." *Annalen der Physik* 309(3):353–363.

Einstein's paper on the photoelectic effect was one of three remarkable publications of 1905.

Einstein, A. 1905. "Concerning an heuristic point of view towards the emission and transformation of light." *Annelen der Physik* 17:132–148.

A description of the curious relationship between Wolfgang Pauli, Carl Jung, and the fine structure constant is provided in

Miller, Arthur I. 2010. *137: Jung, Pauli, and the Pursuit of a Scientific Obsession.* New York: Norton.

The apparent coincidences of fundamental constants that make the universe habitable are described in

Rees, Martin. 2000. *Just Six Numbers: The Deep Forces that Shape the Universe.* New York: Basic Books, Perseus Book Group.

A recent comment on the atomic bomb project with access to recently revealed documents is

Rose, Paul. 2002. *Heisenberg and the Nazi Atomic Bomb.* Oakland: University of California Press.

Many do not share the opinions of Paul Rose.

Chapter 7. Materials and Devices

There are two books that describe a wide variety of real-life physics phenomena. First is

Walker, Jearl. 2007. *Flying Circus of Physics.* New York: Wiley.

This book has many tough questions. The original 1975 edition is especially challenging because it does not have the answers. The second book is not as difficult:

Jargodzki, Christopher and Potter, Franklin. 2001. *Mad about Physics.* New York: Wiley.

A novel in which a fictitious phase transition plays a starring role is

Vonnegut, Kurt. 1963. *Cat's Cradle.* New York: Random House, Dial Press.
Moore, Gordon. 1965. "Cramming more components in integrated circuits." *Electronics Magazine* 38(8): 114–117.

The complexity of device physics is described in many good, but challenging, books. A relatively approachable example is provided in

Roulston, David J. 1999. *An Introduction to the Physics of Semiconductor Devices.* New York: Oxford University Press.

A wide variety of devices, including radio, microwave ovens, lasers, compact discs and much more is described in

Bloomfield, Louis A. 2006. *How Things Work*. New York: Wiley.

Chapter 8. Relativity

Criticism of Einstein that still honors his genius is in

Ohanian, Hans. 2008. *Einstein's Mistakes: The Human Failings of Genius*. New York: Norton.

Generally recognized as the finest book on Einstein and his time is

Pais, Abraham. 2005. *Subtle Is the Lord: The Science and Life of Albert Einstein*. New York: Oxford University Press.

The entropy and evaporation of black holes is described in

Dvali, Georgi. 2015. "Quantum black holes." *Physics Today* 68(1):38–43.

The camel and the eye of a needle is described in the bible:

Matthew 19:24. *New King James*. 2008. Nashville, TN: Thomas Nelson.

Perhaps Albert Einstein never made the famous "biggest blunder" statement about the cosmological constant. The quotation may have been invented by George Gamow, who had an unusual sense of humor.

It is really hard to be sure that the universe is isotropic and homogeneous on the largest length scales. Thus the cosmological principle is frequently challenged by both philosophers and observers of the cosmos. A resolution of this controversy is unlikely to come soon.

Chapter 10. Particle Physics

A good idea about the nature of advanced quantum theory is provided in

Feynman, Richard. 1985. *QED: The Strange Theory of Light and Matter*. Princeton, NJ: Princeton University Press.

Probably the most popular of several books designed to impress us with the exotic world of particle physics is

Greene, Brian. 2003. *The Elegant Universe: Superstrings, Hidden Dimensions and the Quest for the Ultimate Theory*. New York: Norton.

Chapter 11. Nobel Laureates

Steven Weinberg's negative comment about religion is from a conference talk.

Conference on Cosmic Design. American Association for the Advancement of Science. 1999. Washington DC.

A copy of the Humanist Manifesto can be obtained from the American Humanist Society, 177 T Street NW Washington, DC 9009-7125.

Heisenberg's famous quote on religion appears to originate from an article published twelve years after his death.

Hildebrand, Ulrich. 1988. Ethos (die Zeitschrift fur die ganze Familie), No. 10, October. Berneck Switzerland:Swengeler-Verlag AG.

The first citation of Rutherford's comment on physics and stamp collecting appeared twenty-five years after his death.

Birks, J.B. 1963. *Rutherford at Manchester* p 108. New York: Benjamin.

De Broglie's description of how he made his discovery was in the Preface to the re-edited 1924 Ph.D. thesis.

Richard Feynman's comment relating the philosophy of science to ornithology has produced many refutations (many of them pompous). Despite its popularity, there seems to be no source for this comment and it, too, may be manufactured even though it sounds like something Feynman would say.

One suspects that some of the most famous quotes of great scientists may have been manufactured by someone else. The same possibility applies to Einstein's "greatest blunder" comment.

Sudarshan's complaint was reported by

Mehta, Neha. 2007. "Physicist Cries Foul over Nobel Prize." *Hindustan Times* April 4.

Background on the interesting Abdus Salam is provided in

Fraser, Gordon. 2008. *Cosmic Anger: Abdus Salam—the First Muslim Nobel Scientist.* Oxford, UK: Oxford University Press.

The claim that the Nobel committee considered Edison and Tesla for a Nobel Prize is discussed in

Clark, Ronald W. 1977. *Edison: The Man Who Made the Future.* New York: Putnam.

Three, of many, books that survey the lives of famous physicists are

Weber, Robert L. 1988. *Pioneers of Science, Nobel Prize Winners in Physics.* Briston, UK: Hilger.
James, Ioan. 2004. *Remarkable Physicists, from Galileo to Yukawa.* Cambridge, UK: Cambridge University Press.
Dardo, Mauro. 2004. *Nobel Laureates and Twentieth-Century Physics.* Cambridge, UK: Cambridge University Press.

The first of these gives brief and fairly complimentary outlines of every Nobel Prize winner. The second includes carefully researched essays on some of the most famous. The third has a broader scope.

Interviews that provide insight into the personalities of Nobel Prize winners and other famous scientists are in

Hargittai, Magdolna and Istvan. 2004. *Candid Science IV: Conversations with Famous Physicists.* London: Imperial College Press.

Lots of stories, mostly relating to the early years of quantum theory, are in

Capri, Anton Z. 2011. *Quips, Quotes and Quanta.* 2nd edition. Hackensack, NJ: World Scientific.

Maxwell's comment that Her Majesty's government will someday tax electricity is on page 23. Einstein's music stories are on page 38, but in a slightly different form and the musician is reported to be Gregor Piatigorsky. Rutherford's statement "We have no money, so we have to think." is on page 76. Planck's comment to his son about the importance of his work is on page 65. The meeting between Max Planck and Hitler is described slightly differently in this book than in the following reference.

There are differing accounts of the meeting of Max Planck and Adolph Hitler. Planck's quote is from

Ball, Philip. 2014. *Serving the Reich*, p. 62. Chicago: University of Chicago Press.

A description of Pauli's biting criticism is in

Peierls, R. 1960. "Wolfgang Pauly, 1905–1958." *Biographical Memoirs of Fellows of the Royal Society* 5:186.

Index of Scientists, Inventors, and Innovators

Subject Index

Note: Page numbers ending in "f" refer to figures. Page numbers ending in "t" refer to tables.